T0363105

Fungi
of
Aotearoa

An earth tongue (left) alongside two waxgills.

Liv Sisson

photography by Paula Vigus

A curious forager's field guide

Fungi of Aotearoa

PENGUIN BOOKS

PENGUIN

UK | USA | Canada | Ireland | Australia
India | New Zealand | South Africa | China

Penguin is an imprint of the Penguin Random House group of companies, whose addresses can be found at global.penguinrandomhouse.com.

Penguin
Random House
New Zealand

10 9 8 7

Design by Carla Sy © Penguin Random House New Zealand
Illustrations by Carla Sy, unless otherwise credited
Front cover image: Werewere-kōkako by Paula Vigus
Back cover image: Shiitake mushrooms by Tatjana Zlatkovic, www.stocksy.com
Prepress by Soar Communications Group
Printed and bound in China by Toppan Leefung Printing Limited

A catalogue record for this book is available from the National Library of New Zealand.

ISBN 978-1-76104-787-9
eISBN 978-1-76104-788-6

penguin.co.nz

To Bud, my grandmother,
who was also fond of fungi.

Contents

Introduction

The fungi of Aotearoa are fascinating, freaky and fantastical. We have a powdery white fungus that hunts bugs. A basket-shaped species that can move around. A lichen named after Jacinda Ardern. And a blue mushroom on our $50 note. We have others that glow in the dark, and a few that can kill you, liquefy your liver or send you to outer space. And these are just the ones we know about.

Like our flora and fauna, the fungi of Aotearoa have evolved in isolation. They feature brilliant hues, alien textures and unique personalities that often can't be found anywhere else. Each fungus has a story to tell, and this book is a collection of those stories.

My first real encounter with fungi happened in 2015, when I was visiting Aotearoa on exchange. While tramping on Rakiura Stewart Island, I was struck by lichening. Not lightning . . . but lichening (like-en-ing). The weather was terrible. We tramped with hoods up, heads down — and that's when it happened. For the very first time, I noticed the magnificence of lichen — that crusty stuff that grows on every deck, fence-post and footpath.

Lichens on Rakiura range from bubblegum pink to lime green. Some are frilly, others are crumbly. They dangle from branches, cover boulders like splashes of paint, and no two are the same. Each is its own planet of whorling colours and strange topographies. During the three days I spent in the Rakiura bush, I became enchanted by the lichens I saw, their odd textures, the way they clung to nearly everything, and their tiny hairs — visible only if you zoom in as far as your eyeballs will allow. I filled my phone to the brim with photos of it all.

Canadian naturalist Trevor Goward coined the phrase 'lichening rod effect' to describe this kind of moment. By this he means that lichen has a habit of shattering familiar concepts to make way for supercharged understanding and enthusiasm. I thought I knew what forests contained — flora and fauna. But lichens showed me I'd only scratched the surface. They opened my eyes to the 'third F' — fungi. Lichens have had a similar effect on the scientific world, too. Science often looks at plants and animals as distinct entities, individuals that live in constant competition with one another in a dog-eat-dog, survival-of-the-fittest kinda way. Lichens turn that idea on its head.

For starters, a lichen isn't just one thing. It's made up of two main partners, one fungi and one algae. The algae partner provides, using photosynthesis to make food. The fungi partner supports and protects, weaving itself around the algae to make a home. Apart, the two can't survive. But together, they can hack life just about anywhere — even in hostile settings like volcanic vents.

This way of living just doesn't fit the 'everyone for themself' narrative. In 1877, German botanist Albert Frank coined the word 'symbiosis' to describe the relationship that creates a lichen. His lichening rod moment revealed that new language, and new ways of thinking, were required to describe the way a lichen lives.

This way of thinking, though, wasn't as new as Albert might have thought. In many ways, science is simply catching up to what indigenous peoples have known for centuries. As I've learned, an understanding that there are no true individuals is part of te ao Māori, the Māori worldview. Through this lens, we're all connected in a web of life, where a delicate balance of competition *and* cooperation always exists. If fungi have taught me anything, it's that there is never just one way to 'know' something. Science offers us one framework. Within it, we are subjects, while plants, animals and fungi are objects. We are thinking, spiritual beings; and they are matter. But to me, the lichens on Rakiura felt very much alive. They sang out to me with their variety, their vitality, and their voracity to cover every single tree in sight. The two meanings — mine and that of science — didn't match. In te ao Māori, I later learned, everything within the web of life, inanimate or animate, is imbued with mauri, a living essence or spirit. When I look at fungi in this way, I see enchanting beings, with big personalities and unique ways of living. They are truly alive. And their lives are wildly entangled with ours.

I left Rakiura with an eye for lichen, and from that trip forward I saw them literally everywhere I went. Lichen covers 7% of the Earth's surface; it even grows on my 1995 Isuzu Bighorn. When I left Aotearoa and returned to my university in Virginia, USA, lichen still held my attention. I sketched them in my art classes and collected tiny samples during my geology field work. I borrowed an electron microscope to look at them as closely as possible.

One day, while walking down the street I grew up on — a route I'd taken hundreds of times — I got struck by lichening again. This time, what I noticed was *Amanita muscaria*, the classic red and white toadstool. How, I thought, could something so wild, so whimsical, so strange, exist in real life? And could it have been there all along?

In that instant, my childhood home — the most familiar place in my world — became mysterious; an unexplored dimension was suddenly on my doorstep. I returned to this same spot almost daily. I got to know red, orange, brown, yellow, green, blue and purple fungi. All were fascinating. How had I spent 20 years not noticing these characters?

In the midst of becoming 'bemushroomed', I met a Kiwi, fell in love, returned to Aotearoa and kept falling head-first into my fascination with fungi.

Within these pages you'll meet some of the most

charismatic, delicious, shocking and useful fungi that can be found in Aotearoa. The identification tips and foraging frameworks will help you embark on a journey into their world, which is vast, diverse and mostly unknown. Mycologists (fungi scientists) aren't in agreement on how many fungi species are out there. Guesses range from 2 to 5 million, but as one Kiwi mycologist put it to me, 'Our best estimate is . . . a shit tonne.' Some species are mushrooms, but the majority aren't; they're lichens, moulds and mildews, and most, like yeasts, are microscopic. At this very moment fungi are in you, on you, floating by on the breeze, and living in the soil beneath your feet.

Fungi are a requirement of life. Without fungi, Speight's would be just water and sugar. We wouldn't be able to eat plants. Marmite wouldn't exist. And neither would modern medicine. I can't think of a topic that fungi doesn't somehow sponsor. On this journey, fungi have become my teachers. They have filled my days with colour and wonder. They dot the landscape of my memory, and even pop up in my dreams.

New information is constantly emerging about our fungi: we're only just beginning to understand these magnificent organisms. As I've gotten to know them, I've foraged for fungi themselves, but for their stories and teachings, too. This book is a tangle of their tales, and the info you'll find here comes from many sources — data, studies, kōrero, my own adventures and observations. I've gathered all these pieces together into something that is, hopefully, sweet and surprising. I hope this book delights you and jump-starts your own story with fungi. I can't wait to see the meaning *you* make along the way.

Liv's lichen drawings.

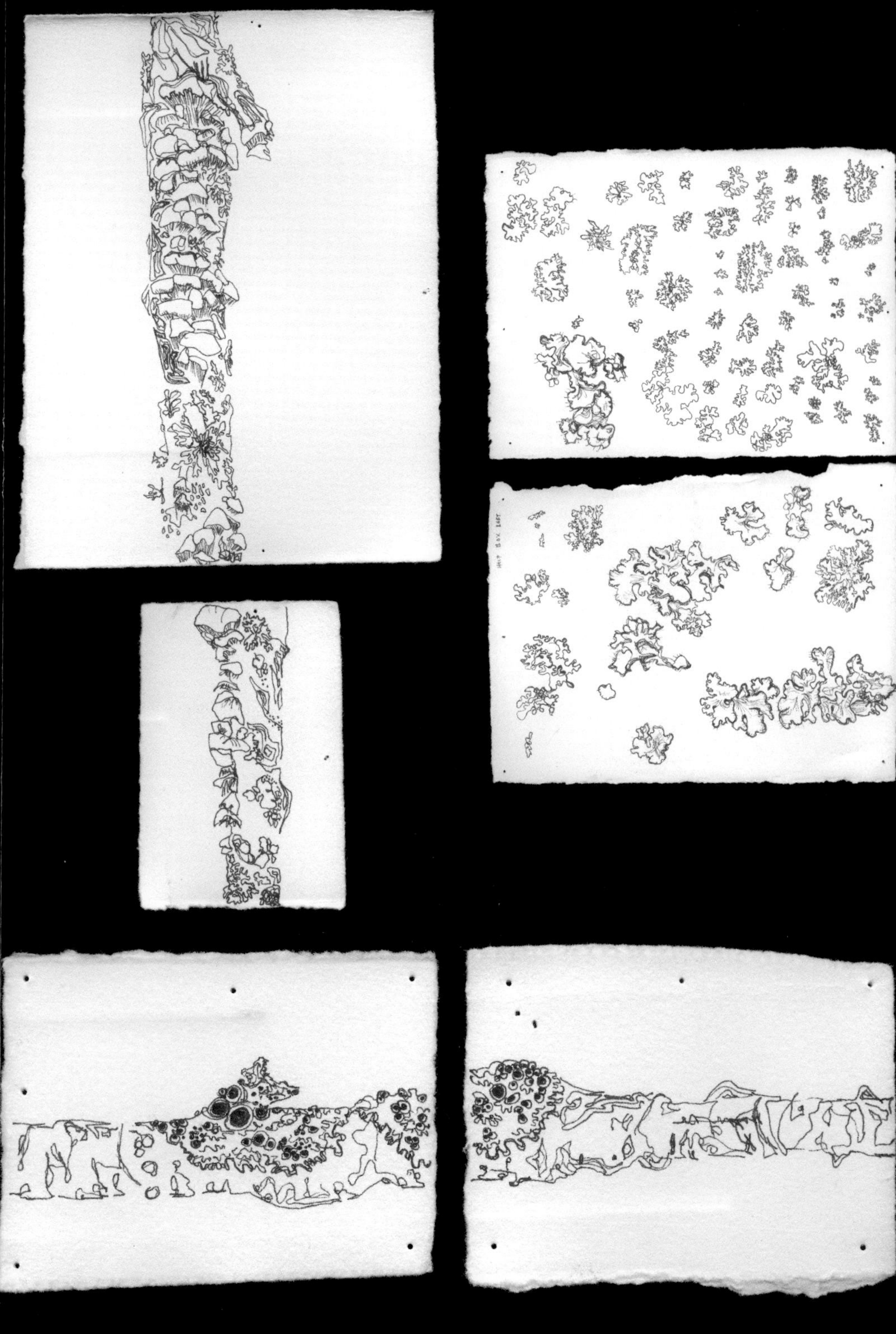

1

What are fungi?

Fungi are essential to life. To understand what fungi are we need to go back in time, to an era that was supremely mucky.

We're 360 million years back, to be exact. It's the Carboniferous Period. Aotearoa is part of the ancient supercontinent Gondwanaland; we're still attached to Australia. On the ground, things are similarly scary. A 2-metre millipede slithers by. Dragonflies the size of toddlers buzz about. Amphibians are everywhere, and the whole place is a swamp. Dead plants pile up left, right and centre. It stinks. I mean, it *reeks*. And it's all lignin's fault.

Lignin is a super-sturdy complex polymer that all vascular plants contain. It helps plants 'stand', but basically refuses to break down. Towards the end of the Carboniferous Period, though, something extremely important happened. White rot fungus evolved a special enzyme that was capable of degrading tough plant material. In short, the white rot clean-up crew showed up to the Carboniferous party and said 'The show is over.' Even now, fungi are pretty much the only major organisms that can break lignin down. Fungi have helped engineer the landscape we know today — one where plants get quickly churned to soil instead of hanging around stinking for ages.

This ground-breaking (plant-breaking?) adaptation also helped make people possible. Without the aid of the microscopic fungi that live in our guts, we wouldn't be able to digest lignin-rich plants or produce the minerals we need to live. And, of course, we also eat fungi. On toast,

Previous page Pixie's parasol, *Mycena interrupta*.
Left A native morel.

atop pizza and in pastas. But mushrooms are only the tip of the iceberg — they're less than 10% of it. The vast majority of fungi species do not produce mushrooms.

So, what are fungi? Mushrooms, moulds, mildews, rusts, smuts and yeasts are all fungi. Different to plants, fungi have chitin in their cell walls, the same stuff that gives crustaceans their hard shells. Fungi, like animals, are heterotrophs — they 'eat' by absorbing dissolved molecules. Fungi secrete digestive enzymes onto the organic material that surrounds them, often wood, to access these snacks. Plants, on the other hand, are autotrophs. They make their own food via photosynthesis. Fungi 'move' in two ways — by growing and releasing reproductive spores, which are spread by the wind.

Fungi are everywhere. At this very moment, microscopic fungi are living in you and on you. You likely breathed in a few fungal spores in the time it took you to read this line. Fungi colonise our bodies, run our ecosystems, and make an absolute mess of the tidy categories we use to organise the world around us.

In terms of size, fungi somehow manage to be both micro and macro at the same time. The largest living thing (by area) is a fungal network in Oregon, USA. It's about 2400 years old and covers 965 hectares, which is about 1665 rugby fields or roughly 10 square kilometres.[1] The blue whale hardly stood a chance against this thing.

Minuscule fungi, the little guys we can only see through a microscope, have been changing the course of history since the Carboniferous. Not only do the fungi in our guts help us digest food, but they also help trigger our development and protect us from disease. And when disease does strike, we have fungi to thank for the medicines that heal us. Penicillin, the original antibiotic, is derived from the *Penicillium* fungi. The discovery of penicillin is credited with helping the Allied forces win World War I, by preventing deaths due to infection. And the human discovery of yeast fermentation set us on a culinary pathway that was unthinkable when hunting, gathering and consuming immediately was the only option. Fermentation and food preservation freed up our time and allowed us to pursue other things like art, music and footy. Thanks, yeast.

Fungi are essential to life itself, and they pave the way for new life too. They're ecosystem pioneers. When retreating glaciers leave the land bare, when volcanoes encapsulate entire ecosystems in lava, or when fires scorch the earth — fungi are often the first organisms to re-emerge. And it's not just any old fungi that lead the charge, it's lichen. That symbiotic collaboration between fungi, algae and, as we've learned more recently, bacteria too. Slowly but surely, lichens colonise rock faces. The acids that the lichens secrete etch away at those rocks, digesting them into the tiniest bits possible. This is the first step in soil production, it releases life-giving minerals back into the ecosystem so that other flora, fauna and fungi can set up shop once again.

Favolaschia cyatheae.

Ordering infinite chaos

If fungi had a LinkedIn profile, their list of skills would be the envy of every CEO. Unlike LinkedIn, though, we don't know how many individuals are part of the network. It's nearly impossible to pinpoint how many fungi species are out there.

Scientists love to debate this topic. One estimate suggests that there are 2.2 to 3.8 *million* fungi species on this planet. Six to ten times more than plants. Based on these numbers, we've only described about 6% of what's out there.[2] If we keep going at this rate, it'll take ages to describe them all.

New species are constantly being identified and described by both mycologists and fungi nerds alike. This happens more often than you might think. At the time of writing, Paula Vigus, whose incredible fungi photographs are featured throughout this book, found an undescribed species while out walking, which will now be officially researched, described and named.

I mean . . . imagine that! In a world where everything feels like it's been done, every avenue feels explored, these waters are more than 90% uncharted.

So even if you're just getting started with this topic, it doesn't matter. Amateurs are welcome. In fact, they can make a huge impact here. By adding photos of the fungi you find to citizen scientist websites like iNaturalist, you can help catalogue what lives in Aotearoa.

Until recently, fungi were scientifically lumped into the same category as plants. For most of human history, we've categorised the natural world only by what we can see — ourselves, animals

Favolaschia claudopus in their multitudes.

that can move around, and rooted plants that can't. Naturally, fungi fell into the anchored, stationary category. In 1751, Carl Linnaeus, who developed the scientific taxonomic method for naming organisms, wrote: 'The order of Fungi is still Chaos, a scandal of art, no botanist knowing what is a Species and what is a Variety.'[3]

The consensus on what fungi are and what they do is only recent. Ancient Egyptians believed that mushrooms were plants of immortality, gifts from the god Osiris.[4] German folklore holds that mushrooms growing in circular ring patterns must be dance floors for witches.[5] There are over 35 different names for our native basket fungus in te reo Māori. One of these names, tūtae kēhua, infers that the fungus might be ghost droppings. Other names include whatitiri, meaning thunder. Maybe loud cracks startled these strange fungi up from below?

The notion that fungi needed a separate kingdom goes back to the start of the twentieth century. It didn't officially get one until 1969, though, when ecologist Robert Whittaker formally named the kingdom according to the rules of Linnean zoological nomenclature. Plants, he noted, 'eat' light and turn it into carbohydrates through photosynthesis. Fungi don't do this. They've got to find their own kai, and that actually makes them more similar to us than to plants.[6] This vein of conversation gets even weirder when you consider that the fungi kingdom has more genetic similarities to the animal kingdom (us) than it does to the plant kingdom.[7]

Fungi are . . . creators and destroyers

We can't live without the fungi in our guts, and we also can't live without the fungi in our soils. About a quarter of all living species can be found in soil — a single teaspoon of it is said to contain more organisms than there are humans on this planet. Fungi make up anywhere from a third to half of the living mass in soil.[8]

The fungi, bacteria and other microorganisms in the dirt are a key part of the system that feeds us. Like the sun and the rain, fungi play a huge part in helping delicious fruit and veg come true. Fungi actually supply up to 80% of the nutrients that plants require.[9] We can't grow food without fungi.

Healthy soils pass on their nutrients to produce, and then on to us. Soils that are teeming with life produce the most nutritious and delicious food. Micro-fungi are crucial contributors to our food system. But their working conditions aren't great: intensive-farming practices like ploughing and the application of chemical fertilisers and fungicides are exhausting these tiny labourers. Monoculture cropping, where one plant is repeatedly grown on the same land, strips the life out of our soil faster than Mother Nature can replenish it. Every year, more fertiliser must be added to the soil ecosystem. Up to 90% of the world's soil could be degraded by 2050.[10]

The flow-on effects of this can be seen in our food. The carrots we eat today, for example, are less nutritious than the carrots our grandparents ate. They also contain less sugar, which means they're less delicious. Today we must eat a larger volume of fruit and veg to hit our daily nutritional requirements. As the life in our soil diminishes, the life in her fruits does too. I saw this play out in real time after several summers of planting corn in the same spot in my garden. My third summer's corn was hungry-as. It required a lot more fertiliser and wasn't nearly as sweet as the first crop.

While critical to life itself, fungi can wreak havoc on it too. Fungi have caused more than a few famines throughout history. Myrtle rust, an imported rust fungus, is threatening our native trees here in Aotearoa, and an elusive fungus-like organism is attacking our oldest stands of kauri. Every year facial eczema, a fungal infection that affects livestock, costs our agriculture industry millions.

Bananas might be the best example we have of fungi as a double-edged sword. When humans first came across wild bananas in Southeast Asia, they were full of hard seeds. Over thousands of years, bananas with favourable qualities were selected for and cultivated; we ended up with hundreds of delicious species scattered across the globe. One of these, the Gros Michel, had become the world favourite by the early 1900s and was grown on huge plantations. It was big, creamy and seedless. Without seeds, though, new plants had to be propagated from the cuttings of another, meaning every plant was a genetic clone of just a few original individuals.

Enter Panama 1, a soil fungus deadly to the Gros Michel. Because each plant was effectively the same, Panama 1 was a threat to all. By 1950, the fungus had spread across the globe and nearly wiped out the Gros Michel. In response, we bred the Cavendish, a similar banana that had resistance to the fungus. Unfortunately, a new fungal pathogen arrived. Panama 4 flourished thanks to rising temperatures, and attacked the Cavendish.[11]

New Zealanders eat about 18 kg of bananas per year, per person. They're the fruit we eat most, right at home sliced over a bowl of Weet-Bix. But they are under threat. Cultivating and eating a range of genetically diverse foods across the world — many types of banana instead of just one — helps beneficial soil fungi flourish and puts a handbrake on the harmful ones. Luckily, there are more than a few New Zealanders invested in ensuring that the future isn't banana-less. They're scientists, researchers, farmers, chefs who are looking to both the past and the future to diversify what we *could* eat and to protect the food resources we already know and love.

Diversifying what we eat

Fungi are sort of like fish. There are a lot of fish in the sea, but just five species make up about 80% of the fish we consume. There are over 2000 edible fungi species globally. And yet, we pretty much only eat one, the white button mushroom. And we're actually eating less variety than we think, too — portobellos are just button mushrooms that have been allowed to grow up. They're the same species.

What we eat as a human community is only a small percentage of what we could eat. And what we do eat is becoming more similar and less diverse. This decreases the resiliency of our food system overall, puts pressure on the micro-fungi in our soil, and opens the door for fungal pathogens to wipe out commercial crops grown as monocultures.

Now, what if we wanted to go beyond the fungi basics? What could we eat then? There are over a dozen edible fungi that grow right here in Aotearoa. In the bush, on your block, even in our CBDs. And if you're thinking 'Nope, I don't like mushrooms', that's fair enough. Many of us know them only as slimy sides served up to us as kids. But maybe you just haven't found the edible fungi and format you like yet.

Legendary Wellington chef and forager Joe McLeod (Ngāi Tūhoe) loves to forage for fungi. They were used extensively in the early Māori pantry, he told me, for food, survival and fire-carrying. Some species like hakeke or woodear were used mostly in lean times; but others, like pekepeke-kiore, might have been seen as more desirable taste-wise.[12] Pekepeke-kiore has a cauliflower-like texture and tastes a bit like crab. It can be fried up fritter-style and goes well with a side of aioli. This species and others that have been, and still are, used as food can be found in the edible fungi section of this book.

A delicious porcini mushroom.

Nutritional powerhouses

Mushrooms give meat products a run for their money. Some mushrooms contain up to 40% protein by weight. They have low (or no) cholesterol, are sodium-free, contain no gluten, are low in fat and sugar, yet still offer us loads of vitamins and minerals. Like plants, fungi are rich in fibre; but they also contain vitamins and minerals that plants are generally low on.[13] Fungi are also a decent source of protein for caloric cost, usually providing about 3 to 4 grams of protein per 100 grams of fungi. For comparison, carrots have about 0.1 grams of protein per 100 grams.

Mushrooms also contain the amino acids we need to use that protein, along with iron, vitamin B12 and vitamin D — all of which can be hard to get if you're trying to eat less meat. Tawaka, for example, one of our native edible fungi, contains up to 20% protein by weight. Tawaka burgers, anyone?[14]

Fungi also regenerate super-quickly. While the tomatoes in my garden take about six months to grow, and it takes yonks to raise a cow, mushrooms double in size almost every 24 hours. Alternative proteins are making waves in the food space, and it does seem to take a lot less resource to raise a kilogram of mushroom protein than a kilogram of animal protein. When prepared right, both are pretty yum, and both are pretty nutritious.

Comparing the nutritional profile of commercially produced veggie burgers and fungi is another interesting exercise. Have you ever looked at the ingredients list on a veggie burger?

Field mushrooms.

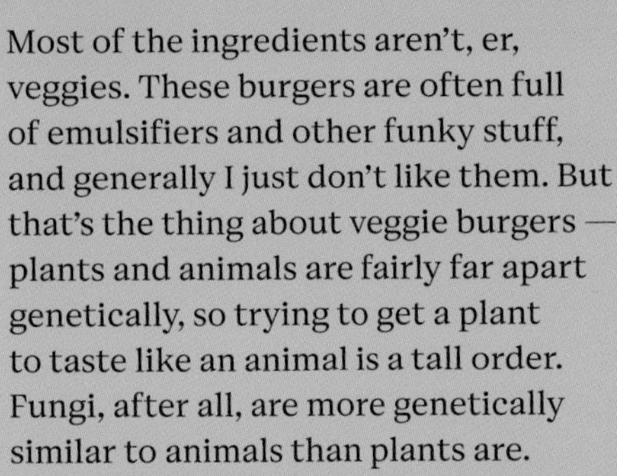

Most of the ingredients aren't, er, veggies. These burgers are often full of emulsifiers and other funky stuff, and generally I just don't like them. But that's the thing about veggie burgers — plants and animals are fairly far apart genetically, so trying to get a plant to taste like an animal is a tall order. Fungi, after all, are more genetically similar to animals than plants are.

Maybe it's not that we need meat 'alternatives', but that we just need more 'meaty' options. A wider variety of delicious foods that can supply us with the protein and other goodness that animal products do. And mushrooms are delicious in their own right. They are simply themselves, rather than trying to be something else or an alternative. Alternatives aren't very helpful if they don't taste nice.

Mushrooms, however, naturally have an umami profile. We can do all these interesting things with them, but it's not magic. It's real mushrooms. That are really meaty.

Fungi are . . . our collaborators

In kōrero with Joe McLeod of Ngāi Tūhoe, I learned about the purple, blue, yellow, orange, white, black, green, even translucent fungi he finds on the ngāhere (forest) floor in his home, Te Urewera. He keeps an eye on the moss and lichen there, too. All are indicators of air clarity, water purity and weather patterns. Joe sees fungi as a telekinetic liquid system that links to the underworld; a legacy of Papatūānuku, who uses it to communicate with plant matter worldwide. Joe taught me that Māori have always been aware of the unique importance of fungi to the stability of our flora and fauna, our food system, and their spiritual connectivity to each other. When out in the bush, he is constantly looking at the whole system as opposed to one small piece.

Along with intensive agriculture and land development, climate change is also putting pressure on our food system, testing the strength of the network. In the bush, the mushrooms that Joe forages seem bigger and puffier than they used to; this could be their response to warmer temperatures and higher humidity. Now what if we looked at our food system as just that, a system? My garden invited me to. When my corn crop didn't grow as well as I expected it would, I wanted to know why. I'd had great success with it for two years in the same spot. So off to the library I went . . .

I found several answers in *Te Mahi Oneone Hua Parakore: A Māori Soil Sovereignty and Wellbeing Handbook.*[15] Corn, I learned, requires nutrient-rich soil, soil that is very much alive. And after the corn is harvested, the soil and the micro-fungi within it need time to replenish themselves. I also learned that in te ao Māori, soil is a taonga; it is the sacred treasure at the foundation of our interconnected world. Everything is seen as part of an interwoven family — animals, fungi, birds, insects, people, soils, the sun. The term 'tangata whenua' — people of the land — reflects this.

This showed me that I wasn't the masterful manager of the garden I thought I was — instead, I was a wishful co-producer, working alongside corn, alongside the soil and the micro-fungi within it.[16] With this in mind, the fact that I'd repeatedly planted corn in the same spot felt silly-as. It was time for a crop rotation; time to enjoy the memory of last year's corn, and plant beans in her place. Beans are nitrogen-fixers — they capture nitrogen from the air and draw it down into the soil. This replenishes our fungal friends after the hard work of bringing a corn crop to life is done.

With this collaborative approach — a balance of giving and taking — the path towards a resilient food system feels more straightforward. It may not be forward, in fact, but back — back to a way of raising food in a way that doesn't forget about our key collaborators. And fungi may help show us the way. There's a lot of hope in our soils and fungi. The more we consider them, the more we might learn about how to do things just a little bit differently.

Fungi are . . . the original social networkers

Most of us first learned about fungi in primary school. For me, the primary school paradigm went like this: *plants produce, animals consume, fungi recycle*. Fungi are the all-important middlemen. They sit somewhere in between life and death, they turn the new into old, and the old into new. But that's only part of their story.

Fungi have evolved more than one way to 'feed' on organic matter (see pages 30–31). They're clever about getting the nutrients they need to live. And diving into how they eat will not only help us understand what fungi are, but will also take us below deck, into the soil, between the tree roots. The underground lives of fungi are nothing short of spectacular.

How fungi eat

Here are the main ways fungi eat. These help explain both what fungi are and what they do.

Decomposition

Saprophytic fungi, also called 'saprobes', are the classic decomposers.

These fungi are the great recyclers of our planet. They break down dead stuff and turn it into usable nutrients and rich soils. They grow on dead or dying organic matter like leaf piles, old wood and distressed trees.

Parasitism

Parasitic fungi grow from a host. The fungi siphon off usable nutrients from the host, or even use the host itself for nutrition, sometimes killing it.

These species are fascinating, spooky, and bad news if you're a beetle, an ant or a spider — lots of insects and animals are plagued by parasitic fungi. You may have even experienced it yourself. Ever had, for example . . . athlete's foot? Fungal pathogens that affect plants, like Myrtle rust, cause disease and also harm their hosts.

Mutualism

Fungi that practise mutualism form a relationship with a plant to secure food. And you've actually already learned about a very famous mutualistic relationship in this book — lichen.

- *Lichens* are a marriage primarily between a fungus and a photosynthesising algae. The fungus offers protection, the algae offers food.

Mycelium growing through a woody substrate.

And together, the lichen can live in environments that the two individuals could not hack on their own.

- *Endophytic fungi* grow within plant tissue. Every plant species contains at least one of these that is critical to its survival.[17] Sort of like how we couldn't live without the fungi in our own guts/microbiomes. In exchange for the plant providing a 'house', the fungi help the plant absorb nutrients and resist disease.
- *Mycorrhizal fungi* have underground networks called mycelium (more on this on page 32). These thread-like networks interact with plant roots to source food. The plant supplies sugars from photosynthesis, the fungus provides nutrients from the soil, and together the two make it work.

Fungi are hard to put in a box, though. They rarely use just one of the feeding strategies described here. Their relationships are difficult to define and are almost never purely cooperative or purely competitive. Sometimes fungi share, sometimes they steal. Sometimes they help, other times they harm. These relationships ebb and flow. Collaboration, at the end of the day, actually requires cooperation *and* competition.

Magnificent mycelium

To understand mycelium, we need to travel through time and space. So hold on to your gumboots — we're going from the garden to the edge of the galaxy here.

If you do work in your backyard, you've probably already seen mycelium. It's white and branching, delicate yet strong, like a fungal spider web, and can be seen in clods of dirt, under old logs, and tucked beneath leaf litter. The individual wisps of mycelium are called hyphae. Each one is thinner than a human hair.

Mycorrhizal fungi and mycelium are the side of the fungi kingdom most of us didn't learn about in primary school. But it's in this subterranean realm where things get extra-interesting. The sheer amount of mycelium in our soil is mind-bending. As is its importance to life.

Mycorrhizal mycelium is the most widespread organism in soil. If each bit of mycelium in the top 10 centimetres of soil were lined up in some strange fungal conga line, it would stretch more than 450 quadrillion kilometres. That's literally half the width of our galaxy. Just one tablespoon of soil contains about 12 kilometres of mycelium. I can hardly wrap my head around that, or even run that far.

Mycelium isn't just something there's a lot of, though — it's also an ancient, life-giving support network.[18] Over 90% of plants have relationships with mycorrhizal fungi.[19] Without these fungal collaborators, plants would struggle to get enough to eat and survive.

At first glance, this doesn't seem to make sense. Plants can make their own food with the sun's rays; surely they're independent individuals? But while plants don't appear to need fungi now, back in the day they seriously did.

Around 500 million years ago, aquatic plants slowly made their way out of the ocean. Once on land they began to evolve their own root systems, a very arduous task. In the meantime, they relied on fungi in the soil to function in place of roots. Long story short: plants and fungi never made a clean break. They're still entangled in crucial, complicated, sometimes petty, and even long-distance relationships.

The role of underground fungi in the storage of carbon is also often forgotten. In this gnarly climate crisis we have on our hands, fungi are a key ally in limiting global heating. When fungi dismantle dead wood, for example, the carbon that was locked up in that wood shifts into the soil. Our underground ecosystems store up to 75% of all terrestrial carbon, which is three times more than the amount of carbon stored in living animals and plants.[20] Globally, underground mycelium draws down, or 'sequesters', up to 5 billion tonnes of carbon dioxide each year. That's roughly equal to the United States' annual CO_2 emissions, hidden away in wispy tendrils of hyphae.[21] Carbon can remain stored there for millennia, or can be quickly released back to the atmosphere. Precipitation levels, vegetation levels, soil texture and land use all affect how long carbon can be stored underground.

As a carbon sink, mycelium plays a crucial role in balancing the global carbon cycle. But as we disrupt and degrade underground fungal networks, through agriculture and land development, their ability to store carbon diminishes. Working with the land in a way that increases its ability to hold carbon could be a key step in our race against climate collapse. Less tilling, more cover-cropping and more crop rotation are all strategies that support mycelium's ability to keep storing carbon and limit global heating. A 0.1% increase in underground storage of carbon would remove 100 million cars-worth of carbon from the atmosphere.[22]

The plant and fungi relationship

In simple terms, here's how mycorrhizal relationships work between fungi and plants.

Mycelium winds and weaves its way through the soil, foraging for nutrients as it goes along. Its wisps of hyphae grow around and through plant roots and offer up their gathered resources, like nitrogen and phosphorus. But mycelium doesn't just do this for free. There's an exchange to be made, a deal to be struck.

Photosynthesis is what plants bring to the table. They take in sunlight and use it to turn water and carbon dioxide into sugars, which they then send down into their roots. The mycelial threads within and around those roots suck out some of the sugars to feed themselves.

Even though plants have now evolved their own root systems, they're still dependent on their fungi allies for nitrogen, phosphorus and other goodies from the soil. But can't trees, for example, pull that stuff up with their own roots? They can — but most trees can't 'mine' enough nutrition from the soil to survive. Their roots, compared to fine hyphal threads, are bulky-as. While tree roots stumble around looking for tasty morsels, the nimble mycelium wanders with ease, finding its way into even the tiniest pockets of soil space. In this way, mycelium massively increases the surface area accessible to the tree's root system and helps it gather vastly more food.

So it's not that fungi *can't* photosynthesise, it's that they don't need to. And even though plants can photosynthesise, they still need fungi to survive. Fungi are a foundational part of every landscape we've ever known. Even the most familiar forests have subterranean stories and lives we're only just beginning to understand.

An occasional mushroom

Mushrooms are the frontmen of the fungi world. When we hear the word 'fungi', they're what we think of first. But remember, only about 10% of fungi species produce mushrooms. And they only do so when it suits them. Most fungi species are microscopic, like yeasts. So, what *is* a mushroom? Where does it fit into the fungi story?

A mushroom is like a fruit, the fruit of a fungus. When the temperature and moisture levels are right, the millions of hyphal strands that make up a mycelium will expand, swell, knit together, and produce . . . a mushroom! But why? What's the point of it? Like a periscope peeking out from below, the mushroom connects the fungus to the world above.

The mushroom represents one of the ways some fungi spread. It contains reproductive spores, which, once mature, will be forcibly ejected and spread by the wind. The science is complicated, but, briefly, each spore contains genetic information. If it meets with a sexually compatible spore of the same species, it could establish a new node of the fungus. The probability of this is low, but mushrooms can release over a billion spores per day. As the mycelium expands underground, this helps the fungi spread too. Some fungi produce fruiting bodies with bright colours or other enticing attributes. Some truffles, for example, are highly aromatic. This beckons animals to come in for a nibble, which in turn spreads the spores.

The humble apple tree provides a helpful but imperfect analogy for what a mushroom is and what it does. The tree is always around, just doing its thing. When the conditions are right, though, it will produce apples. The apples that fall to the ground might turn into smaller saplings. The apples that get scavenged might spread that apple tree species to new areas. Picking an apple does not harm or kill the tree. Just as plucking a mushroom does not kill the mycelium. With fungi, we usually can't see the 'tree'. But it's most certainly always there.

The other 90% of fungi species are mostly too small to see, but you actually know them pretty well already. Microscopic fungi, along with bacteria, play a key role both in growing our food and preserving it. From salami to sauerkraut, from kombucha to kvass to kimchi, from miso to yoghurt, plus tempeh, ketchup, soy sauce and more, almost every culture on Earth has key fermented items in their cuisine. These fermented and preserved foods sit at the heart of many food stories and have played an almost unknowable role in freeing up our time, keeping us well, and bringing us together around the table.

The forest redefined

In 2016, I spent some time researching beavers for a uni project with the US Forest Service. I surveyed our research site, inventoried the trees, took water quality samples and played with curious swallowtail butterflies. I never once saw a beaver.

One thing I *did* see, though, was what looked like a bundle of white asparagus growing straight out of the ground. It was spooky, unlike any plant I'd ever seen. I made my way out of that dark, cool section of the forest, back to the research van, with the feeling that I'd happened upon something special. Back at the lab, I looked it up — it's called ghost pipe, *Monotropa uniflora*. We don't have this exact species in Aotearoa, but we do have some interesting equivalents, like *Gastrodia sesamoides*, the potato orchid.

Monotropa has no leaves, and it grows in dense parts of the forest where almost no other plants can. It's not too bothered about the light situation because it can't photosynthesise — it doesn't contain any chlorophyll, the chemical that gives most plants their green colour and allows them to turn light into energy. No leaves and no photosynthesis? These are the hallmark features of plants, yet *Monotropa* doesn't have, or need, either. How, then, does it survive?

In the late 1800s, a Russian botanist proposed that maybe *Monotropa* drew essential nutrients from underground fungi. This idea did not catch on in the slightest. But, although not proven until 1960, it was right.

***Monotropa uniflora*, ghost pipe.**

This ghostly plant, I learned after spotting it, is a mycoheterotroph. Mycoheterotrophs parasitise fungi. *Monotropa*, for example, siphons carbon from underground mycelial networks to get the nutrients it needs to live. Those fungal networks, of course, generally get their carbon from other plants via their mycorrhizal relationships.

Almost all plants rely on mycorrhizal fungi for some amount of nutrition. Within these relationships, there's an exchange. Fungi supply plants with foraged soil nutrients and plants supply fungi with sugars produced with the help of the sun. But mycoheterotrophic *Monotropa* doesn't photosynthesise. It relies on fungi for breakfast, lunch and dinner. And it does not offer its fungi food source anything in return.

The scientific, Darwinian way of knowing the world paints a picture of many distinct individuals all living in constant competition for survival. Mycorrhizal fungi shatter this idea — showing us that amidst the competition, there's also cooperation. *Monotropa*, though, highlights that relationships are rarely a 50:50 split. This oddball plant sent me on the learning journey I'm still on. It turned a stone for me, and underneath that stone I've found a few surprising stories. Two of my favourites are the Wood Wide Web and the Mycelial Market.

The Wood Wide Web

The strange, pale plant *Monotropa uniflora* shows us that nutrients can move from fungi to plant and from plant to plant via fungi. Almost no one has contributed more to our understanding of this than Suzanne Simard, a forest ecologist at the University of British Columbia. In 1997 she published a fascinating scientific study that illustrated just how complicated and intriguing relationships on the forest floor can be.

In the study, Simard exposed birch seedlings to radioactive carbon dioxide. Two years later she returned to the same patch of woodland and saw that the carbon had passed from the birch seedlings to nearby fir seedlings, which shared a mycorrhizal fungal network — but not to nearby cedars, which did not share the network. And when the fir seedlings were shaded, and their ability to photosynthesise was therefore limited, they received more carbon from the birch tree 'donors' than unshaded firs did. Simard showed that within the forest system, carbon moves from places of abundance to places of scarcity or need. This work challenges the idea that plants are distinct entities that only compete with one another.[23] It suggests that rather than focusing on competing individuals, we need to look at the whole community to understand how a forest 'works'.

Information can also be shared via fungal networks. Simard and her colleagues planted Douglas fir seedlings and ponderosa pine seedlings next to one another, with mesh barriers in between.[24] The mesh prevented the trees' roots from touching but allowed enough space for contact via mycorrhizal mycelium. The researchers then stressed out the fir seedlings by pulling off all their needles. The naked trees then sent a 'warning' signal through the fungal network to the pines, which produced protective enzymes in response.

Simard noted that fungal networks operate like neural networks in our brains, or even the internet. They move resources and information to places where they're needed, when they're needed. On the back of these studies, Simard coined the phrase 'Wood Wide Web' to refer to this sophisticated system that sits at the foundation of our forests.

Simard has also helped to paint a picture of just how buzzy and intertwined the world beneath our feet is — this world we cannot see. In a 30 square metre stand of Douglas fir trees, she mapped out exactly which trees were connected to one another. The most well-connected tree — the 'mother tree' — was linked to 47 others via two species of fungi. The most well-connected fungi had 19 connections.[25]

According to Simard, mother trees 'look after' seedlings in their network, providing them with extra nutrients as they struggle upwards from the noisy and hyper-competitive understorey. When these great old trees, the keeper of their system's 'memory', begin their long, slow death, they can dump their nutrients into the system and share their remaining resources preferentially with younger trees who are their offspring.[26]

Bird's nest fungi.

The Mycelial Market

The term Wood Wide Web is catchy and inspiring, but fungi do get a bit lost in there. Within the Wood Wide Web framework, fungi sometimes feel like a passive partner; just a pipeline. But if you look at things from the perspective of the fungi, things get really interesting. This is where the idea of the Mycelial Market comes in.

The Mycelial Market shows us another way to 'know' what mycelium is and what it does. Through this lens, fungi are not just silent middlemen — they are brokers on the forest floor who oversee deals big, small, good, bad, fair and unfair. Their relationships with trees are complicated — it's not one fungus and one tree, sharing equally. Far from it.

Toby Kiers, an evolutionary biologist at VU Amsterdam, sees the Wood Wide Web as a market economy rather than a social welfare system. In one experiment, she saw that fungi could dictate the 'price' of phosphorus. Kiers exposed one mycelial network to two phosphorus supplies — one plentiful, the other scarce. In the plentiful area, the 'price' that trees had to pay for the phosphorus was lower — let's say one unit of tree sugar for one unit of fungi phosphorus. Where the phosphorus was in shorter supply, though, the trees had to make a stronger offer, like two sugars for one phosphorus. The fungal network was also able to transport phosphorus to the area with the better exchange rate, to capitalise on demand.[27] Sound a bit like your Economy 101 course?

I'm not sure which mycelial analogy is correct, the apple tree, Wood Wide Web or the Mycelial Market, or any of the others we could choose from. But maybe we don't need an analogy. Mycelial networks are simply, distinctly, extraordinarily . . . themselves. However you want to think of them, they've up-ended our scientific ideas of what forests do and how they live. They've reminded us that it's not as simple as survival of the fittest.

Fungi also challenge our concept of the individual. Is a

lichen two individuals, one fungus and one algae, living side by side? Or is it one big individual? And what do you make of yourself? When you consider the millions of tiny fungi living within and on you, are you still 'just you'? You couldn't live without them. The line is blurry. Very blurry.

While the conversations around mother trees and mycelium sometimes feel ground-breaking and new, the understanding that we're all connected, in one big web, isn't new at all. Many indigenous cultures have understood for centuries that there are no true individuals, that all life from flora to fauna to fungi is inherently one.

Fungi are . . . medicine for the planet

Fungi may offer our planet modern remedies for modern problems, like single-use packaging. In Aotearoa, about 252,000 tonnes of plastic end up in landfill every year.[28] Polystyrene is even worse. It takes up loads of space in landfill, often gets blown into waterways, and takes 500 years to break down.

BioFab in Auckland *grows* a polystyrene alternative using fungi. It's called Mushroom™ Packaging, although technically it's mycelium packaging. Packaging made from mushrooms isn't that many steps away from how mycelium behaves and functions in nature. To make it, woodchips are placed in a mould, fungi spawn is added and the resulting mycelial growth binds the woodchips together to create the shape desired. In terms of strength, mycelium-based material and polystyrene are right on a par with one another.

BioFab's partner in the USA, Ecovative, has even built a house out of mycelium. Such a house might have some unique benefits here in Aotearoa. We are short of more than a few homes, and building supplies can sometimes be very hard to get hold of. But what if we could build houses from hyphae? Could we grow cladding, or insulation? Mycelial building materials aren't yet in a form where they meet New Zealand building codes, but they have huge potential and the technology is evolving quickly.

Packaging made from mushrooms sounds like an exciting start, but what if it gets too popular, too fast? Compostable packaging is great, but a lot of it requires energy-intensive commercial composting facilities. And those facilities increasingly do not want compostable packaging — it doesn't add any value to the compost and there's too much of it for them to process. So off to landfill your 'compostable' coffee cup goes. But that isn't likely to happen with packaging made from mushrooms. It's biodegradable in the way you probably think it should be. You can tear it up and add it to your compost heap in your back garden. It also provides nutrients to whatever system it's added to; as the mycelium breaks down, the nutrients it contains become available to microbes in the soil.

Running out of woodchips, the base material, also shouldn't be a problem as the packaging process can use almost any plant biomass as a base. You could use weeds with high biomass, like dock, which grows on every fence line in the country. Or even hemp, after the seeds have been harvested. Hemp grows huge in just two months. There's great potential to create an entire industry around locally sourced packaging made from mushrooms — the process uses relatively low energy, only requires a bit of spawn to get started, and there are plenty of forestry and agricultural hubs here that could provide the growing medium.

Aotearoa also has an overload of wilding pines.

These trees were introduced in the 1880s to combat hillside erosion and were later planted in huge plantations to create our forestry industry. Unfortunately, they've escaped from those original plantings and have become a super-spreading invasive species. It's estimated that if their spread isn't managed, 25% of the country's landmass could be covered by wilding pines within 30 years. Wilding pines already cover more than 1.8 million hectares of Aotearoa. Despite efforts to control them, they have continued to spread by about 5% a year. That's about 90,000 more hectares blanketed per annum.[29]

Currently, the two main methods for managing wilding pines are felling the trees and spraying them. These methods are expensive, have their own negative impacts, and aren't 100% effective — even after an area is treated, the surrounding soil will still contain pine seeds and will produce new saplings within a few years. Fungi researcher Genevieve Early is investigating how we might be able to use native fungus harore (*Armillaria novae-zelandiae*) to infect the pines and kill them naturally, to slow their spread and protect our native bush from their advance.

Armillaria is an impressive decomposer. The idea behind Genevieve's research is to inoculate wilding pines with the fungus and let it get to work. This ecological method of control can hopefully be used alongside the current methods of management to achieve longer-term suppression of wilding pines.

One great thing about this strategy is that it works with what we already have here. *Armillaria* is a native species and spreads quickly. It helps keep things tidy — without prolific decomposers like this one, we'd be chin-deep in dead wood. Its current role in our ecosystems seems to signal that using this fungus to control wilding pines won't pose a risk to native bush since it's already an integral part of that system.

When we make a mess of the environment, fungi can help us restore balance. In 2006, for example, researchers in New Zealand showed that white rot fungi could be used to efficiently clear the pesticide PCP from contaminated soils.[30] Home gardeners can even get involved with this sort of thing. There's lots of good research showing that 'myco-restoration', or the introduction of certain fungi spawn to soils, can help pull out and break down icky stuff and restore healthy soil ecosystems. Several companies in Aotearoa sell the spawn and the woodchips you need to get started with this.

Fungi are . . . medicine for us

Fungi may help us heal the Earth, reduce our impact on her, and could be a key collaborator in another, more intimate environment too — our own bodies.

Superbugs are scary-as. They resist almost every kind of antibiotic, and they're on the rise in Aotearoa as well as the rest of the world. They cause staph infections and other gnarly health complications, and by some estimates could kill more frequently than cancer by 2050. But there's good news: fungi may be able to fight superbugs.

Fungi are one of the foundations of modern medicine. Penicillin, the first mass-produced antibiotic, came from *Penicillium* fungi. Antibiotics alone have added 23 years to the average human life expectancy.[31] Before we had them, one in every nine skin infections led to death.[32]

Penicillin ushered in a roaring period of antibiotic discovery. This period produced drugs like cyclosporin, which suppresses the immune system and makes things like organ transplants possible. Cyclosporin was found when a pharmaceutical company encouraged employees to bring soil samples from their holidays back to the lab, to see if they contained any useful fungal strains.[33]

Fungi have actually been used as medicine for a lot longer than antibiotics have been around. Māori used fungi as medicine as well as kai. Pūtawa were cut into strips and used to dress wounds. Puffballs were used to stop bleeding and to treat burns and scalds. Tawaka was given to expectant mothers and to people suffering from fevers. It was also used to treat karaka and tutu poisonings. Angiangi and other lichens were collected and used to dress wounds and slow bleeding.[34] In China, fungi have been used as medicine for thousands of years. Medicinal mushrooms first appeared in Taoist art around 1400 AD, around the same time that the first Polynesian peoples may have arrived in Aotearoa.[35]

Black earth tongues (top) growing alongside waxgill mushrooms (bottom).

Superbugs, however, pose a scary question — are we in a post-antibiotic era? Antibiotic-resistant superbugs, like MRSA, have become more resistant and harder to fight each year. In 2014 the World Health Organization warned that within a decade, antibiotic-resistant bacteria could make routine surgeries really risky and write the end of modern medicine. In Aotearoa, our geographical isolation has offered a slight superbug buffer. But in 2009, the first case of a carbapenem-resistant organism (CRO) was identified here. Carbapenems are a high-powered group of antibiotics used to treat infections that other drugs can't. Every year, more patients with CROs are identified in Aotearoa.

Our current antibiotics have got us this far, but we need new ones. The good news is that there's still loads of potentially useful fungi out there — we just haven't had enough time to get to know them all yet. But we're working on it. Siouxsie Wiles and her bright pink locks graced our TV screens almost daily during the Covid lockdowns. She kept us up to date on the facts, breaking down the jargon so that Kiwis everywhere could better understand what the heck was going on. Since 2015, Siouxsie has been working with Bevan Weir to investigate local fungi for potential antibiotic applications. Our flora and fauna are wildly different compared with those in the rest of the world, and the same is true of our fungi. What we have here is unique, and could hold what we need to make some new medications.

Importantly, we also have the International Collection of Microorganisms from Plants (ICMP). Located in Auckland, this collection holds 23,000 living cultures — strains of fungi, bacteria and other microorganisms stored frozen in tanks of liquid nitrogen. This collection is the product of 70-plus years of work and is one of the best in the world. It contains more than 10,000 fungi species from Aotearoa and all over the Pacific and could well be an untapped trove of useful compounds. In one of Siouxsie's lab's early findings, 35 out of 36 fungi samples from the collection showed some kind of activity against mycobacteria, a family of superbugs that includes *Mycobacterium tuberculosis*, which causes TB. The lab has also identified a fungal strain with some activity against the hospital superbug methicillin-resistant *Staphylococcus aureus*, better known as MRSA.[36]

So what does this mean? Have we got a new wonder drug,

a penicillin 2.0? Not quite. The scientific process is rigorous and thorough, and so it's normally quite slow. Identifying potential candidates is just step one, with many steps between that and actually using something in a drug trial. Our Covid-19 vaccines were developed quickly, but only on the back of decades of research into mRNA vaccines.

Now, why do fungi 'work' for us as medicine? To start with, fungi can't 'run away' from threats like pathogens, so they have evolved clever counter-strategies — an entire arsenal of chemical responses to fight off whatever bacteria or other nasties come their way. And their defence strategies often work really well for us too. But why?

At the core of things, we're not that different. We are genetically quite similar to fungi and are pestered by many of the same viruses.[37] Fungi produce many chemical compounds to protect themselves, but the best understood are the beta-glucans, which have been shown to stimulate weak immune systems.[38] Many of the most treasured mushrooms in traditional medicine, like *Ganoderma* species, are high in these compounds.[39]

Our shared history with fungi is another key part of this puzzle. Animals, including humans, that have been able to identify and use fungi as food and medicine have gained physiological benefits in the short term. And evolutionary benefits in the long term. Along the way we have evolved to have receptor sites where we can process and use what fungi provide. As a result, when we consume medicinal fungi they trigger healing, nourishment, defence — or all three.

Fungi are . . . magic

Every so often, magic mushroom spores end up in a council woodchipper. Once that woodchip is spread, warm rain coaxes curious fungi into being on library lawns and along police station sidewalks. These fungal blooms fire up furtive Facebook groups and bring out a different kind of forager. Magic mushrooms contain psilocybin, a mind-altering (psychoactive) substance that is a Class A drug in Aotearoa. It's illegal, as are the mushrooms themselves. They pose a high risk of harm to humans.

So how does this happen on council's watch? There isn't anything dodgy going on — magic mushrooms grow up and down Aotearoa, all of their own accord. We even have a few native and endemic species. Their spores are often dispersed by the wind and, when those spores come to land on woody debris, and that woody debris comes to land in the woodchipper, councils can end up being co-conspirators in the spread.

Humans and other animals are known to seek out altered states of mind. Kererū, our native wood pigeons, eat fermented miro berries and sometimes fall out of trees when they overindulge. Kids spin around to get dizzy. Adults use caffeine to dial in, and alcohol to wind down. In his book *How to Change Your Mind*, American author Michael Pollan writes, 'There is not a culture on earth that doesn't make use of certain plants to change the contents of the mind whether as a matter of healing, habit, or spiritual practice.'

This got Dr Mitchell Head (Tainui, Ngāti Mahuta, Ngāti Naho) thinking. Did, or does, rongoā Māori — traditional healing — make use of the mind-altering substances that naturally occur in the ngāhere? Were early Māori aware of endemic, psychedelic mushroom species such as *Psilocybe weraroa*? Mitchell is a neuroscientist whose research sits at the intersection of science and mātauranga Māori (Māori knowledge). He's looking into how *Psilocybe weraroa* could be used as a medicinal product to treat addiction in conjunction with a cultural therapeutic framework. This fungus contains psilocybin and is a taonga (sacred) species.

Now, an *illegal* substance . . . that might be used for *therapy*? At first glance it doesn't add up. But there are incredibly strong signals, both current and historical, that substances like psilocybin magic mushrooms can be used as powerful additions to our mental health toolkit. Peyote, a spineless psychedelic cactus, has been used in Mexico and the American south-west for more than 6000 years. The Navajo, a Native American tribe, have used it to treat alcoholism and in ceremonies designed to connect people with spiritual power, guidance and healing. Recent research in Aotearoa and abroad has found that psychedelics could be used to treat depression, addiction, and even the sense of dread that comes with terminal diagnoses. Psychedelics, importantly, are not addictive. And the benefits shown in these studies are often not only immediate, but lasting.

These kinds of applications are what Mitchell is interested in. But how do the benefits 'work'? The 'snow-globe' analogy helps explain this. Imagine going sledding after a fresh fall of snow. At the start of the day, you can choose any path down the hill. But after a few goes, the tracks of the previous runs are bitten into the hillside. After an hour or two, the original tracks are deep and established. Once your sled is on them, it becomes harder and harder to swerve and forge a new path to the bottom. Depression is often fuelled by repetitive ruminative thinking loops; addiction is a repeated behaviour. Both represent deeply rutted thinking patterns that are hard to deviate from.

Experiences, also called 'trips', induced by psychedelics have been likened to 'shaking the snow globe'. When you turn the globe upside down and back again, the snow resettles, and this creates an opportunity for new paths to be taken. A psychedelic experience can disrupt unhealthy patterns of thought and create new, flexible ways of thinking and behaving, even if you're fairly far down a difficult life-track.[40]

The benefits of psychedelics used as therapy, Mitchell taught me, can be seen in the way fungi operate in the natural world — they're great at creating communication pathways. When you put them into the brain, they accelerate the growth of neuronal pathways. Every time we perform a particular behaviour, we create feedback loops and those behaviour pathways get reinforced. Psilocybin allows the brain to follow new pathways rather than staying in the same ruts. It's all about putting energy into the system and allowing it to take a new path.

As many indigenous cultures overseas have a history of using their native psychedelic species, it didn't seem likely to Mitchell that Māori hadn't experimented with theirs, too. Through his research and kōrero, he has found that there is indeed such knowledge but things like the

Tohunga Suppression Act of 1907 forced discussion underground. During his research, Mitchell has found several accounts of traditional Māori use of magic mushrooms, generally passed on through oral tradition. Mitchell is now working with a marae to develop a therapy for their people struggling with addiction. With the help of Manaaki Whenua, the group explored the bush around their marae and found *Psilocybe weraroa*. No one has yet quantified how much psilocybin these mushrooms contain. Mitchell is hoping to not only do that, but also determine a therapeutic dose that can be achieved consistently. The second, and equally essential, part of this research is the cultural framework Mitchell is working on.

Psychedelics are powerful substances. 'Set and setting' are key in producing a generally positive experience. *Set* refers to the mindset the user brings to the experience, while *setting* refers to the physical place in which it occurs. Indigenous practices, like Navajo peyote ceremonies and the Pacific kava ceremony, provide frameworks for set and setting. When psychedelics are used without this understanding, the experience could be quite scary. People should never look to forage their own magic mushrooms, Mitchell told me — more than a few who have tried in Aotearoa have had negative experiences. And, of course, it's illegal.

Currently we don't have a framework for psilocybin use in a Māori cultural setting. When you take these substances as medicine, Mitchell explained, they act as catalysts for whatever kind of process you want to undertake. Defining the set and setting, and the intention behind taking the medicine, is really important. To design a cultural therapy to wrap around therapeutic psilocybin use, Mitchell has looked to other sacred ceremonies held on marae, such as tā moko (traditional tattooing), and is working with tohunga (healers) and cultural therapists in the marae setting who are experts in the spiritual space to develop this therapy.

Fungi are far from being just 'one thing'. The various forms and definitions they take on know literally no bounds. From feast to famine, magic to malady, they have shaped the world as we know it. They have even shaped us. Our shared history with fungi is wild and wonderful. And it's only just the beginning.

The native ngāhere (forest).

2

How to find and identify fungi

I wasn't fungi-obsessed as a kid, but I *was* fascinated by rockpools. On one summer holiday, I simply *had* to find a starfish; I'd never seen one. My siblings and I spent days looking. There were heaps of amazing anemones and bristly barnacles, but no stars. On the last day of our trip, we gave up and went for ice cream on the pier. I leaned against its wooden rails, enjoying my frozen treat. And from this high perch, about three metres above the rockpools, I finally spotted a starfish.

We rushed down to the pools, and suddenly there they were. Hundreds of them. Tiny and colourful, spiny yet whimsical. The invisible was suddenly revealed to us, and the reality of what we could see in those swirling, miniature worlds expanded before our very eyes. We spotted one after another, after another. Almost like they'd played a trick on us, the starfish were suddenly everywhere.

I've carried that moment in my back pocket ever since that day. It taught me that to find starfish, fungi, or any other subject of interest, all you need is a willingness to pay attention, an understanding of how to look, and the occasional ice cream.

What we experienced in the rockpools is called the 'popout effect'. It's the simple concept that we don't truly 'see' everything in front of us at all times. If we did, our brains would quickly become exhausted by the onslaught of detail. The world's colours, textures, sights and sounds would overwhelm us. Focusing on one thing would be literally impossible. What we 'see' instead is a blend of both

Previous page *Mycena roseoflava*. **Top left** *Cruentomycena viscidocruenta* on a pinecone. **Bottom left** *Gliophorus lilacipes*.

reality and memory. Our brain interprets each rockpool with a bit of new information, and fills in the blanks with memories of rockpools seen earlier.

To notice something new — to spot the starfish, the mushroom, or whatever else you're seeking — your retinas require a search image. Once this search image, say that first starfish, has been acquired, it's highly likely that another, and another, will seemingly appear out of nowhere. It's this moment that fungi foragers are referring to when they say 'Ah yep, now I've got my eyes on.'

To get your eyes on and find fungi, all you've got to do is go for a walk. You'll know the moment when it happens, and the ritual of getting your eyes on will become a treasured part of your fungi-spotting practice. When your eyes come on, it feels like a curtain has been pulled away. It's magical, surprising, and joyful. This response makes sense if we think back to our history as hunter-gatherers. When a pattern is recognised and you've found what you're seeking, synapses fire, carrying information to your brain. 'Fungi!' you might think, and occasionally those fungi are also — 'Food!' Looking for something — getting your eyes on — is a way to tap back into this positive feedback loop.

If you don't spot fungi right away on your walk, don't fret. They are there, waiting for you to find them. It might take a while for your eyes to adjust, for you to locate the right search image and reconnect with this ancient skill. When it happens, you'll feel a jolt. A previously nondescript part of your backyard or local park will seem rife with magic. A vacant lot might become super-wondrous. A familiar forest can suddenly feel unexplored. In an instant, even the most recognisable realities, like your own lawn, can become awe-inspiring.

Here are some of the things I've learned along the way when it comes to finding, identifying and getting to know fungi. Use them to get started on your own personal foray into this wondrous world.

To find fungi

Go small

You don't need to travel far or wide to find amazing fungi. They will often find you. In your search for fungi, going small is a great idea, especially if you're just getting started. Work with what you've got and where you're at. Pick a tiny plot of land to start with. Your backyard, a corner of a local park . . . somewhere you can quickly and easily get to. Steal away to this spot for an hour each day, or even just five minutes. Visit as often and as regularly as you can. Watch it change throughout the year.

I did this accidentally during the 2020 Covid-19 lockdowns. Daily walks led me down my street in the Ōtautahi Christchurch CBD and into the Red Zone, the part of the city most badly damaged by the earthquakes. You can still see where the houses used to be, but now this land labelled red is returning to green.

On this small circuit I started to notice new stuff. Field mushrooms. Puffballs. Edible weeds. Sour sumac. Lemony tarata flowers. Heritage apples. Heavily laden quince trees. The occasional porcini mushroom. As I spotted new-to-me fungi and flora I would jump in delight, take loads of pictures on my phone, then rush home to uncover who they were with the help of my growing shelf of field guides. Within a few weeks I was hooked, returning almost every day to be amidst the hum of the Red Zone, which I'd previously thought was pretty drab.

Identify your microclimates

Fungi can grow just about anywhere. I've seen pukurau (puffballs) busting through impossibly small cracks in concrete. And an *Agaricus* species sprouting from a bathroom wall in Karamea. As you start to look for fungi, there are lots of ways to home in on where the species you're keen to meet can be found. Developing this instinct takes a bit of time, but an understanding of microclimates will help you along.

To understand microclimates, think of Aotearoa. The upper North Island generally has a warmer climate, while the South Island is more temperate. Within that general pattern, you'll find tropical weather in Nelson, snowy alpine environments in Arthur's Pass, and rainforests on the West Coast. Aotearoa is a nation of microclimates.

Keep zooming in, and you'll find that climate is almost infinitely divisible. Even a single harakeke plant, for example, can create a tiny microclimate. It provides shade and shelter, keeping the soil beneath it just that bit damper. Fungi can often be found in cool, moist microclimates — along stream banks, under the canopy of old trees, in the bush, or on the shady side of a park. By identifying a few microclimates near you, you'll be able to vary your search for fungi and find a huge array of species without travelling too far from home. Fungal wonders are almost always close at hand.

Even those living in urban areas have plenty of microclimates to choose from. From the Christchurch CBD, there are plenty of microclimates I can access on my bike. In Hagley Park I've got to know introduced fungi that grow in association with old exotic trees. Along the shaded banks of the Ōtākaro Avon River I often find birch boletes. In the Red Zone, I find edible species like the pukurau that love fringe environments and open grassy areas. On Banks Peninsula, I wander the paths and occasionally spot native fungi growing among the tōtara, kānuka and kahikatea trees.

To find a few microclimates near you, 'fly' around using Google Maps. Dark green areas are usually shadier, cooler and damper, meaning they're great for fungi. Lighter areas are hotter and drier. Identify some local microclimates and then go take a look.

The main microclimates in Aotearoa that are great for fungi-spotting are:

- **Native bush:** there are heaps of colourful fungi that grow in association with beech, mānuka, kānuka and other native trees.
- **Grassy pastures:** these are the places to look for field mushrooms, pukurau, and other edible varieties.
- **Urban/developed areas:** these areas tend to have introduced trees which come with interesting, and occasionally delicious, fungi partners.

Get familiar with fungi forms

Once you begin stumbling across cool fungi in various microclimates, you'll quickly require a new language to describe what makes one fungus different to the next. With the right words, you can understand and know fungi in a new way. I learned this in the rockpools too. As they stretched out with starfish sparkle before us, so did other elements of life. Our baby brother started babbling away, in a rush of discovery forming his first full sentences as we explored the pools. Suddenly, he could describe what he was seeing and ask questions about it all. With a command of language, you can begin to know starfish, fungi, and other elements of nature, more deeply too.

Fungi come in all shapes and sizes. Some are mushrooms, but most aren't. Some look like they could be underwater, in a rockpool, or attached to a coral reef. Others look like they've come from outer space. And a few look like they've wandered off the set of a horror movie.

The macro fungi we have here in Aotearoa — the big fungi you can see with your own eyes — come in many different forms. The following are a few big groups to get you started.

1

Mushrooms — Mushrooms are classically capped fungi. Sometimes called toadstools, these have a stipe (stem) and a cap which has gills beneath it. This is where the mushroom's reproductive spores are held and released from once they've matured.

2

Boletes — These are the same general shape as a mushroom, but with one key difference: underneath you'll find pores (lots of tiny holes) rather than gills. These holes are the open ends of many tiny tubes. Bolete spores form within the tubes and are forcibly ejected once mature, where they're caught and spread by the wind.

3

Pouches — Pouch fungi are shaped like little drawstring purses. Some have small stipes, others have none at all. They hold their reproductive spores within their bodies, which are fully enclosed. Sometimes pouch fungi are semi-subterranean and will just barely peek out of the dirt.

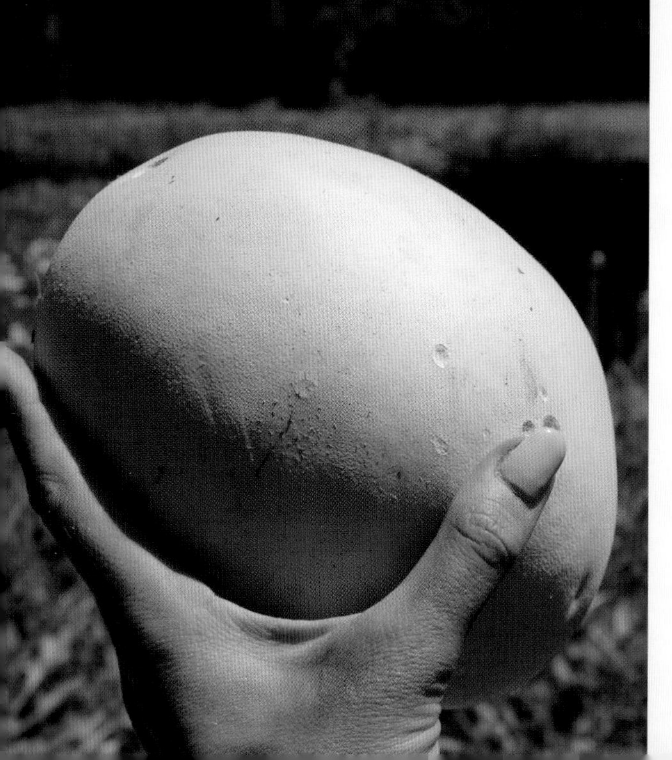

4

Puffballs — These fungi are round and sometimes squishy. They come in neutral colours ranging from buff to brown, and grow from the ground. They don't have stipes, but sometimes you can see a few tangly threads at their base that attach them to their mycelium underground. Puffballs hold their spores within their bodies, which become dry and dusty when mature.

5

Birds' nests — These ones look like tiny birds' nests. They have a cup-like shape which contains 'eggs', which are little parcels of spores called peridioles. When raindrops splash these from the nest, this helps the fungi spread.

6

Stinkhorns — This group is funky in more ways than one. Stinkhorns grow from the ground, have a phallic, anemone or basket-like shape, and are quite smelly. Their spores are held in a sticky, stinky goo which attracts flies and other bugs to help them spread.

7

Corals and clubs — These fungi stand upright. Many have beautiful branching structures that look like underwater coral. Others stand solo and have a club-like shape. Corals and clubs come in all sorts of colours.

8

Cup fungi — Cup fungi are shaped exactly the way the name suggests. Microscopic sac-like structures inside each cup hold the spores in rows within. Some cup fungi are small and delicate; others are large. Some have stubby stipes; others are taller.

9

Jelly fungi — These fungi have a jiggly, jelly-like texture and often grow on wood. They come in a wide array of colours and can shrivel up during dry spells, but then rapidly expand back out once the rain returns.

10

Bracket fungi — Bracket fungi are generally large, dry and have tiny pores on the underside that release the spores. These fungi tend to grow from wood. Some have a hard, shelf-like form, others are more flexible and leathery, and some form a simple crust.

Lichens — Lichens can grow on almost any surface. Some are completely two-dimensional and look like a splash of paint. Others are leafier, and some are dramatically 3-D with upright branching fruiting bodies.

Insect killers — Parasitic fungi that attack insects can sometimes form dusty crusts on their victims' exoskeletons. Other insect-killing fungi stalk insects in the soil, colonise their bodies, and then send up tall, stalky, spore-bearing growths from the insect's remains.

Candle snuff fungus, *Xylaria hypoxylon*.

Parts of a mushroom

Mushrooms are the fungi we know best and eat often. Getting to know the parts of a mushroom will help you describe what you find, and make identifications. This is especially important if you're looking for edible fungi, which we'll dive into next.

1 **Cap** — The fleshy top part of the mushroom, also called the pileus.
2 **Warts/scales** — When some mushrooms first emerge, their juvenile forms are completely encased in a whitish covering called a universal veil. As the mushroom begins to grow, the cap pushes through the universal veil and sprouts upwards. Bits of the veil sometimes remain on the cap in the form of warts or scales.
3 **Hymenophore** — This is the surface where the mushroom's reproductive spores are held. Hymenophores can be gills (as seen here), but they can also be smooth, toothy, or pored.
4 **Spores** — Spores are one way mushrooms reproduce. Once the spores are mature, they are forcibly ejected from the hymenophore and spread by the wind. If one spore meets with another sexually compatible spore of the same species, it could create a new node of the network somewhere else.
5 **Stipe** — Plants have stems, mushrooms have stipes.
6 **Skirt** — When some mushrooms emerge, their gills are covered by a partial veil. As the cap opens more fully, the partial veil falls away and hangs from the stipe as a skirt. Also called an annulus or ring.
7 **Volva (egg sac)** — Some mushrooms have a bulbous, egg-sac-like structure at the base.
8 **Mycelium** — Below the mushroom is its mycelial network, which is made up of thousands of thin strands of hyphae. When the conditions are right, the network swells and knits together to produce a mushroom on the surface which is made of the same material.

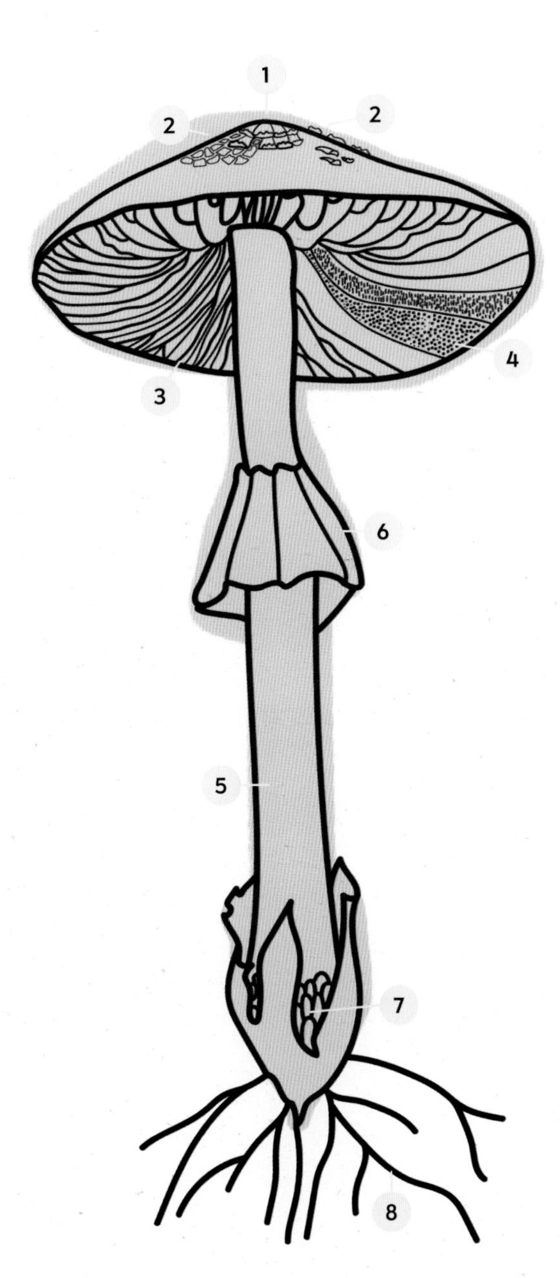

1	Cap
2	Warts/scales
3	Hymenophore
4	Spores
5	Stipe
6	Skirt
7	Volva (egg sac)
8	Mycelium

To identify fungi

Develop your practice

Fungi bring us right to the edge of our own reality. They transport us away from the noise of life, and into a simpler, yet more mysterious, landscape. My dad learned this as a boy. One summer evening he went camping with a few of his mates. In the middle of the night, they discovered that the ground was glowing all around them. Their camp was shimmering in the darkness. The boys had never seen anything like it, and it gave them the creeps. They grabbed their stuff and ran home, scared by what must have been bioluminescent fungi in the soil.

When we bump into the realm of fungi, when we find that surprising other reality, how do we make sense of it? Identification is a great place to start. Use the tips and frameworks in this section to build your own fungi identification practice. With this in your back pocket, every walk — whether it's up a mountain or to the office — is a chance to add more data to this part of your brain.

First, here's a hot tip. Identification apps are pretty much useless. Snap a picture they say, upload it, and voilà, the app will tell you what you've found. Some of these *sort of* work. But no algorithm can replicate the relationship that begins to form when you get your fungi-spotting eyes on and enter their world.

Go slow

When you spot a fungus, the best thing you can do is slow down. Get on their level — on your hands and knees, even. Engage all your senses except taste. Greet the fungi and let them greet you.

Rather than diving in to pluck a beautiful specimen, slow down. There are lots of areas where disturbing fungi, or the land in general, is not allowed (for more on this see pages 88–91). Instead, listen to all that these miraculous organisms have to tell you. Take note of where they're living and how, notice the others living nearby. Take in smells, colours, shapes. All of this data will help you identify what you've found.

Use the following checklist to guide your observations.

It will help you take a complete snapshot of the fungal world you've stumbled into.

Start with the big stuff

- **Environment:** Where are you? Where is the fungus? Some fungi, as we've learned, are saprobic decomposers — they grow on trees or dead wood. Other species are mycorrhizal symbionts — these grow from the ground, their subterranean mycelium commingled with tree roots. The final group, super-spooky parasitic species, grow from a host plant, insect, animal, or another fungus.
- **Neighbours:** What other organisms surround this fungus? Trees really help here. If you're looking at a mycorrhizal species, what trees are growing nearby? If you've got a saprobe, what type of wood is it growing from?

Begin to zoom in

- **Body:** Are you looking at a classically capped mushroom? Or is it a puffball? A pouch fungus? Or a lichen? What colour and texture does it have? What colour is the interior flesh?
- **Underside:** If the fungus has a cap, what's underneath? Gills? Pores? A toothy texture?
- **Gill attachment:** If the fungus you've found has gills, how do they connect with the stipe? Some gills attach to the stipe, whereas others do not come into contact with it. The different kinds of gill attachments in cross-section are illustrated on page 72.
- **Stipe:** If you've found a capped mushroom, what does its stipe look like? Is it smooth, rough, furry, slimy, hollow, or solid? Does it have a skirt?
- **Base:** Is the base straight or bulbous? Does it emerge from a volva? How does the fungus attach to whatever it's growing from?
- **Spores:** What colour are they? There are a few ways to find out. Start by taking a peek underneath the fungus you've found. Is there a light dusty coating on the grass below, or maybe on shorter specimens? Those are the spores! If not, gently brush your finger along the underside, and see if any spores are deposited.

  If these strategies don't work, take a spore print. Simply place the specimen's cap on a piece of paper,

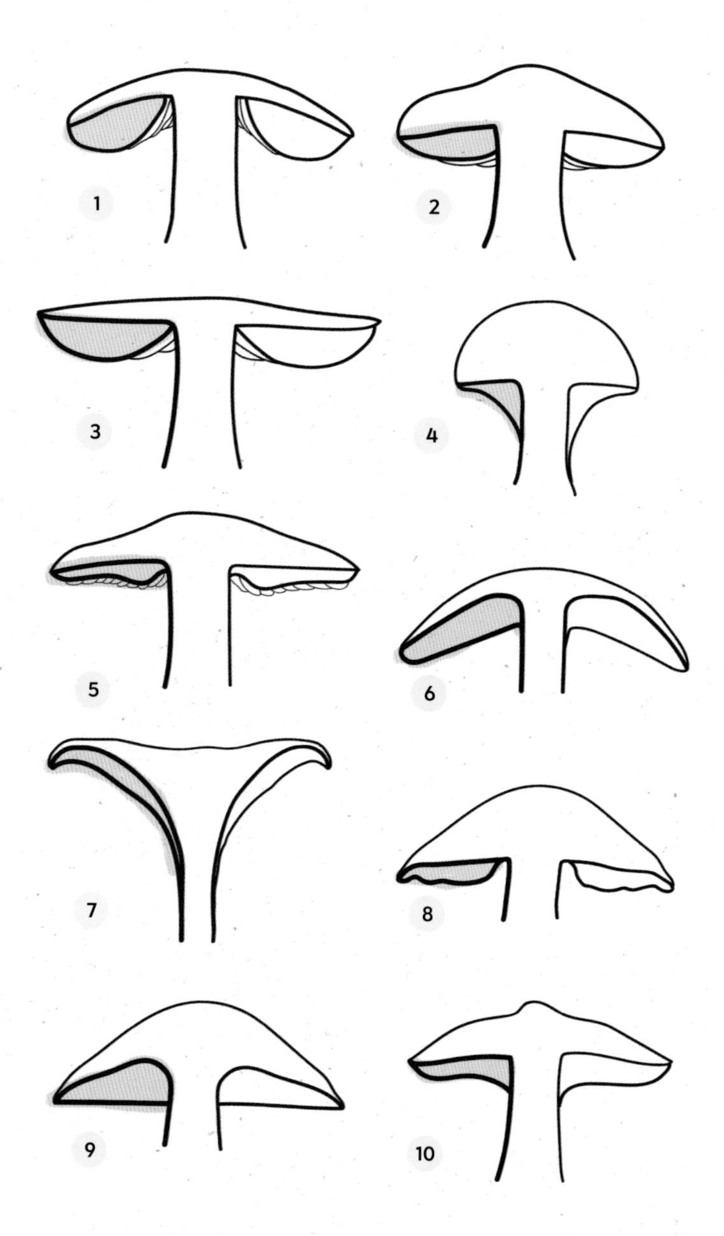

1 Free
2 Emarginate
3 Adnexed
4 Decurrent
5 Sinuate
6 Adnate
7 Arcuate-decurrent
8 Distant
9 Triangular
10 Sub-decurrent

cover it with a bowl and wait a few hours for the spores to drop out. The spores will form a negative image of the gills on the paper below.

- **Staining:** Gently bruise the cap and stipe with your fingernail and watch what happens. Does the exposed flesh change colour? If it does, this is called a stain.
- **Scent:** Some species have distinct aromas, and sometimes field guides reference this. When I first started foraging, I found this odd. Smell feels totally subjective. But here's the thing: as you develop your fungi-spotting practice, you'll begin to notice that you know a particular mushroom's scent. One day you'll smell a mushroom before you spot it, and then you'll realise just how far you've come on this journey. When looking for edible species, bad or acrid smells can also help you identify poisonous fungi.

With this checklist you can gather a huge amount of data even without a microscope, although a hand lens may come in handy! Cross-reference your observations with the field guide sections of this book, and others, to get closer to an identification. This researching and identifying phase is almost as exciting as looking for fungi. Once you know who they are, you can learn their stories too.

Learn their names

Give fungi a little attention and you'll soon be seeing them everywhere. And with words to describe them, they'll start to feel more familiar. Their names will teach you even more about who they are and how they live.

Scientific names, written in the form *Genus species*, are very specific, and help us distinguish one fungus from another at the most granular level. But they don't tell us how to *know* these individuals. The scientific way of speaking doesn't give voice to the alive-ness that fungi have. Science often speaks about nature as things rather than beings. Kids do the opposite. They gender bugs, and name pet rocks. Somewhere along the way to adulthood, we stop doing that.

I'd never thought about this until I learned about 'puhpowee', a word from the Potawatomi Native American language. Puhpowee refers to the specific force that boosts mushrooms up from the soil, and helps them silently punch through the boundary between above and below. I learned about this precious word in *Braiding Sweetgrass* by Robin Wall Kimmerer, a writer, scientist, and member of the Potawatomi Nation. 'Science can be a language of distance,' she writes, 'which reduces a being to its working parts; it is a language of objects . . . The makers of this word however understood a world full of being, full of unseen energies that animate everything.'[1]

Indigenous ways of knowing recognise plants and fungi as non-human beings. When we call them by their names, we recognise them as subjects rather than objects. This creates a space where relationships can form, a space where we might recognise that we are fungi, and fungi are us. While just nine speakers of Wall Kimmerer's language remain, te reo Māori is increasingly used on the news, in business, at school. I'm early on in my reo Māori journey, but fungi help show me the way. When I learn their kupu (vocabulary), I learn about who they are, where they live, and how.

Pukurau, for example, is one of many Māori names for puffballs.[2] Puku means to swell, which is quite fitting for these sometimes huge, orb-like species. Puku can also mean secretly or silently.[3] The first time I came across pukurau, I found 20 all at once, on a walking route I take almost daily. The morning before there had been nothing; but the next day, almost two dozen pukurau, all the size of my head. It felt like they must have rolled in while I was sleeping.

Angiangi, meanwhile, is a species of *Usnea* lichen used in rongoā Māori (traditional Māori medicine). It has powerful antibacterial properties and is found throughout the ngāhere.[4] Angiangi clings to tree branches and trunks, and hangs from them in delicate, light-green wisps. The word angiangi also means a gentle breeze, and angi on its own means to move in a floaty, free manner.

If you can't figure out a fungus's name, or make an identification right away, that's okay. Make some notes about the fungus's personality, what makes it alive. Before long, and maybe after a few more sightings, it may just reveal itself to you.

You now know how to get your eyes on, how to look for fungi, and are beginning to understand how to know them too. I can almost guarantee that you'll soon be seeing fungi everywhere — on your street, at the supermarket, on T-shirts and TV. I can't wait for you to experience the pop out effect; it's truly magical.

A green waxgill species.

How to take photos of fungi

As you come across new-to-you fungi, take some pics of them! Photos will help you identify your finds. This is also a great way to document what you spot in areas where disturbing fungi is not allowed.

When you get home, use your photos to cross-reference with the field guide section of this book. You can also use them to get expert input: share your fungi photos to iNaturalist if you need help getting an accurate ID. Facebook groups like 'Mushroom Hunting New Zealand' are helpful too.

Paula Vigus is an award-winning New Zealand fungi photographer. Her photos are featured throughout this book. Here are her top tips on how to take good fungi photos:

1. **Get on their level** — Take shots from the top, side, and underside if you can manage. Get down on your hands and knees for the best vantage point.
2. **Include context** — Take some snaps of nearby trees and plants. Identifying what grows alongside the fungus can be a very helpful clue.
3. **Take close-ups** — Zoom in on the cap, gills, stipe, and any interesting textures, colours or forms you find.
4. **Use selfie mode for the underside if the angle is tricky** — Put your

phone camera in selfie mode, then place it under the fungus. Focus the camera with a tap, then snap. You'll end up with a hilarious fungi selfie and a snapshot of the underside.

5. **Use flash if you're deep in the bush or if the light is fading** — Or have a companion hold their phone torch over your find. You can even buy a few cheap LED lights to take with you on your fungi-spotting missions.
6. **Take care with focus** — Fuzzy fungi shots aren't helpful later, and fungi identification groups can't work with them. Give yourself a second or two to really get your pics in focus.
7. **To level up your fungi photo game, get a macro lens** — There are magnetic macro lens attachments for phone cameras that are affordable. Macro lenses for real cameras are a big step up, but can seriously capture the visual delight of fungi.
8. **Take heaps!**

3

How to forage for fungi

'Those look kinda gnarly,' is what my partner had to say about some foraged mushrooms I brought home one autumn evening. He was right. Foraging for fungi can be gnarly. In 2020, a New Zealand woman nearly died after misidentifying some wild mushrooms.

There are fungi in Aotearoa that can kill you. But this knowledge should serve as a hefty dose of healthy perspective rather than a complete roadblock. Of the known fungi species, of which there are tens of thousands, only about 3% are known to be poisonous. Within that, there are varying degrees of likely harm. Some will upset your tummy; others will liquefy your liver. About 30 species are consistently fatal; and of those 30, one is responsible for about 90% of mushroom-related deaths worldwide. Its name? The death cap.

Death caps are reportedly quite tasty, and their effects can sometimes take days to show up. They grow in Aotearoa, as a handful of people have learned the hard way. The fungi I'd brought home, though, were porcini. They're near-impossible to confuse with death caps and are actually fairly straightforward to identify. Porcini are unbelievably good. I'd thought they were a European delicacy, so was surprised to find them growing here, literally on my block. We made a brown butter pasta with those precious porcini, and by the end of that meal my partner was on Team Porcini too.

Foraging fungi safely is all about taking the time to read

Previous page A red waxcap mushroom. **Top left** Some inkcap species can be eaten, but this one is used to make ink! **Bottom left** Harvesting a few caps to make ink. Learn how to do this on page 121.

the environment, to read the fungi, and to get a positive identification. An understanding of what to look for, what to watch out for, how to get a positive ID and plenty of patience will keep you out of trouble. Here are some of the best tips and frameworks I've found on how to forage safely and respectfully, and have fun along the way. Even when you don't find anything!

Get a positive ID

There's a big difference between identifying a fungus and actually eating it. When you're foraging for edible fungi, or any other wild kai, you *always* need to get a positive identification before you eat it.

Making a positive ID is all about making decisions not with *total* certainty, but with *enough* certainty that you feel comfortable going ahead. If you want to be 100% sure about the species of the fungus you've found, you'd need to examine its microscopic features or even its genetics. For a hobby forager, this is generally not an option. Positive IDs are key for safe, everyday foraging.

Getting a positive ID is two parts protocol and one part up to you. Here's how to do it:

1. **Observe the fungus you've found and its context.** Document all of its features. What general form does the fungus have? Is it a mushroom, a puffball, or something else?
    — Look at the fungus's smaller features, too. Does it have gills, pores, teeth? Does it have a stipe? Check out every feature you can.
    — Take note of the environment the fungus is living in. Is it growing from the ground or from dead wood? What plants and trees are nearby?
    — Use the checklist on pages 70–73 to capture a complete snapshot of the fungus and its environment.

2. **Cross-reference your observations.** Check what you have observed against trusted field guides, cross-

reference several identification resources, and ask experts for their opinion on platforms like iNaturalist. Use the field guide sections of this book as well as any of the resources listed on page 325.

- If you think you have a porcini, for example, confirm that each and every attribute of it aligns with the identification information in your trusted resources. As soon as one attribute deviates from what you should see, you've got a negative ID. Check your finds against dangerous look-alikes. Do NOT rush through this process.

3 **The last step is all about you.** Only you can decide when you're happy with your sleuthing and confident in your ID. Only you can decide when you've crossed that threshold and know the fungus well enough to cook with it.

- Freaked out by this grey area? I was, too. But we actually make positive IDs all the time. They help us navigate life. Think of crossing the street: you look both ways, take in as much information as you can about the scene, and decide whether or not it's safe to cross.
- The more you forage, the more comfortable you'll become with this practice. Each time you go, you'll get a tiny bit better at finding fungi, and observing them, and you'll hone the research regimen required to land on a positive ID. At some point, the steps to getting a positive (or negative) ID on a fungus will become second-nature. As you get started, stick to the easier to ID edible mushrooms, which are listed in the next section.
- And remember how I said that identification apps are kinda useless? This way of knowing another organism, understanding what makes them who they are, using that understanding to get a positive ID . . . you can't really turn that into an algorithm. It is precious knowledge that can't be pinned down and takes time to gather. It's also worth noting that apps and keys from overseas aren't always applicable here in Aotearoa; our fungi are incredibly unique.

Know the risks

Everything comes with inherent risk. Including foraging for fungi. Eating a poisonous mushroom is a risk you can manage by taking the time to get a positive ID.

Pollution and spraying are other risks that foragers need to think about. Is it safe, for example, to forage a mushroom from the side of the road? Or from a berm that might have been sprayed? In my view, this is up to each and every forager to decide for themselves.

Here's how I think about it. Foraged foods make up a very small but very exciting percentage of what I eat; the rest comes from the supermarket. A huge percentage of our conventionally grown foods are raised with the use of harmful pesticides and chemicals. Like risk, chemicals are inherent in life, and what we do to avoid them is often up to us. So when I happen upon a chunky porcini on a berm or along a footpath, I usually pick it, and cook it up with loads of butter. I feel comfortable with that choice, but it's okay if you don't. If you want to know more about spraying near you, contact your local council. Encourage them to spray less so we can preserve our wild kai resources.

Getting run over by a car or getting lost in the bush — these things can pose a bigger risk to you while foraging than the fungi themselves. Be aware of your surroundings and use your head.

Almost every fungi poisoning that I've heard of has had two things in common. The first is the species: typically the death cap (see page 212). And the second is being in a hurry. Even experienced foragers can get tripped up if they rush to make an identification. While not a complete list by any means, the grid at right highlights species that could be confused with edibles in Aotearoa.

In the species descriptions in the 'Edible fungi' section, I've included notes on dangerous look-alikes and how they are different — but the most important thing is still to **check thoroughly against trusted sources and get that positive ID**.

The more you look, the more you will hone your practice, get comfortable with making IDs and discover the capability that you have to learn within this framework. And while you're learning, you can safely grow your own — see pages 99–100.

Poisonous fungi

1 **Death cap** — could be confused with field mushrooms.

2 **Laughing gym** — not for eating.

3 **Earthball** — could be confused with puffballs.

4 **Yellow-stainer** — could be confused with field mushrooms.

5 **Purple *Cortinarius*** — could be confused with wood blewit.

6 **Funeral bell** — could be confused with velvet shank.

7 **Poison pie** — the name says it all.

Take the right stuff with you

Take the right kit with you to stay safe while you forage. Go well prepared, and you'll have more fun on the journey. Even if you find nothing, you will still have been for a nature walk, and will have done your steps for the day.

1. **Layers** — There's not a lot that can kill you in the outdoors in Aotearoa, except the elements. Take layers.
2. **Water and snacks** — Unlike some fungi species, we cannot survive without these things for very long.
3. **Sunblock, hat, etc.** — Bloody ozone hole. Be sun smart.
4. **Knife** — Good for cutting away slimy, buggy bits of fungi and getting a whole, clean specimen. Get one that folds in on itself for easy and safe carrying.
5. **Basket or container** — Fungi are hardy but delicate, so transport them with care. Bring a basket to satisfy your inner hobbit, or a lidded container if you're going for speed.
6. **Tea towel or pastry brush** — Use this to quickly clean your finds. Tidying them up as you go along will keep them nice for cooking. Add one grotty specimen to the lot and they'll all be grotty by the time you get home.
7. **Field guides** — Bring this book and other guides along with you, for referencing on-the-go.
8. **Phone with a full battery** — For snapping pics of new-to-you species you come across — check these out later to figure out what they might be.

1
2
3
4
5
6
7
8
A PHOTOGRAPHIC GUIDE TO
MUSHROOMS AND OTHER FUNGI OF NEW ZEALAND
GEOFF RIDLEY
PHOTOGRAPHS BY DON HORNE
A Field Gu
to
New Zealand Fun
Shirley Kerr
SONY
BACK COUNTRY
MADE WITH 50% RECYCLED CONTENT

Do the dance

Foraging for fungi might sound intimidating at first. One day, though, you'll realise that you know — I mean really *know* — a certain species on sight. This kind of knowing, gained and gathered over time and through lived experience, is called tacit knowledge. Although even once you get to this level, you still need to confirm your gut feeling and get a positive ID.

Tacit knowledge can't be taught. It's the excitement of seeing an old friend each time you spot a species you can identify from a metre out. This is the knowledge that comes with time and repetition, from forming a relationship with another organism, from a foraging practice that you craft, polish and grow each time you go for a walk.

It can take a wee while to get to this level of knowing, and it's sort of like learning the steps to a dance. You are one partner, and fungi is the other. With each walk or waltz, you learn a bit more and get more comfortable with the steps. You get the hang of anticipating your partner's next move, you understand them better. You can't control the rhythm, but you *can* learn to flow with the seasons and the rains. And along the way you'll learn how others — the weeds, the birds, the fruit trees — flow too.

I sort of just know what field mushrooms are now. I feel joyful every time I find them. When I don't find them, I'm reminded that I have no control over the outcome. In a world that values do-this-get-that transactions, fieldies teach me something different: how to dance with uncertainty.

Forage reciprocally

One of the first things I learned about how to 'be' outside was the 'Leave No Trace' framework. LNT is a simple concept: leave only footprints, take only photos. Foraging is basically the exact opposite to LNT. It's a take. Plain and simple. So it just doesn't fit into the LNT model. When I first started foraging, I felt weird. Who am I to pluck this porcini from the park? Or to gather a bag of walnuts, which the trees have put all their energy into?

Fungi foraging brought up some of these questions for me and made me consider LNT a little bit more. In some ways I think it's pretty reductionist: *If we just leave it alone, it'll be fine. It's better off without us*. Sometimes this is true, but to live we must collaborate with our environment. Foraging has invited me to consider this more intimately, and on a personal level. I *could* take this pretty puffball, or this dinner-plate-sized tawaka, away with me for dinner, but *should* I? If I do take it, what can I offer in return?

As we've learned on this fungi journey, fungi help trees by providing them with nutrients that the trees find hard to access. The fungi don't do this for free, and they remind us that life is a balance of competition and cooperation. So, how do we achieve balance when foraging, or within our relationship with the land generally?

The colonial industrial agricultural model, for example, has enabled us to produce more food than ever before, but it has also stripped our soils of their nutrients, interrupted essential mycorrhizal relationships, and left us with food that is less nutritious than what our grandparents ate. Too much taking, and the land is forced to do less giving. There are examples of this everywhere.

When we *can* balance competition and cooperation, we land on reciprocity. Reciprocity is not about no taking; it's about exchange for mutual benefit. In the space that reciprocity creates, a relationship that promotes care and balance can emerge. To me, that feels a lot more helpful than 'leave it alone'.

There are so many things you can do to forage reciprocally. Maybe you remove litter as you go. Maybe you take only a few edible finds away with you and leave the rest. Maybe you bring a friend along to teach them about the fungi and the land. Maybe you do some reading about what you saw when you get home. You could even join a Manaaki Whenua Landcare Research working day to help with a site restoration.

How to follow the rules

Regarding the LNT framework, though, do be aware that it might not always be clear where you can and can't forage. Plenty of environments in Aotearoa are *not* for foraging — for example areas under rāhui, land protected by legislation, private land, Department of Conservation (DOC) land or fragile environments. Read the signage. Do a bit of research. Check the local council restrictions. Talk to someone who knows the land better than you. Find out who the kaitiaki (guardians) of that area are. And if you're still not sure, listen to your gut feelings. If you shouldn't take something, your gut will tell you 'Don't mess with that.' Foraging pushes me to go beyond LNT and into a partnership with the land around me. Whether it's your own backyard, your street, or your whole rohe (region), foraging can be just that — friendship.

How to always have a fun foray

How do you manage, a friend asked me once, when you go fungi-hunting and find nothing? For me, it's about thinking of it as more of a seeking ritual, rather than a hunt.

As we've learned, you don't need to go far to find enchanting fungi. They're all around us. Looking for them often and in areas you know well is all it takes to kick off your own fungi-spotting practice. By focusing on the seeking part, you'll get the benefits of the activity every single time you do it — even if you find nothing.

It's actually impossible to find *nothing*. Sometimes the most fascinating finds are new observations of old friends. We often think of flora, fauna and fungi as things. But really they are living systems, just like us, through which matter is continually passing. To spot a pine tree or a mushroom is to witness a single moment in its ever-changing life.[1]

Fungi-spotting is a chance to get exercise, sunshine, fresh air. When you happen to find a fungus, it's exciting. It's a natural find, but there's a wow factor every time. There is a thrill to it. When I leave my desk, my screens, and go for just a five-minute walk, if I see a mushroom I still get so excited. The find is its own little universe. And I get to enjoy it for a few moments; detach myself from work for a bit, marvel at it, and then move on. It's like a mini-vacation for free.

Foraging allows us to escape the real world for a bit and enter a smaller, simpler space that is also massively mysterious. It prompts us to re-engage our senses, and brings wonder, awe and joy to even the most mundane moments and spaces. And these aren't just anecdotal benefits, either. This is called behavioural activation: the practice of changing your physical state to improve your mental state. By going for a five-minute foray, by looking, hearing, seeing, touching, and doing something real and away from your technology, you can ground yourself and return to the present moment refreshed.

And when you find something that amazes you and you feel that sense of awe, that's good for you too. Several studies have shown that even after just minutes of staring at magnificent, towering trees, or watching small water droplets dance about, or any other natural wonder, people are more likely to act with greater empathy and kindness in the hours that follow.

Looking for fungi can be a powerful practice, regardless of whether we return home with a haul or not. It's a ritual that can spark wonder, remind us of our place in the web of life, strengthen our connection to the whenua, and reinforce our kinship with her.

Top foraging tips

Follow these top tips from some of the most experienced foragers in Aotearoa, including the legendary Peter Langlands, to keep yourself safe.

For even more information on wild kai, foraging and how to do it right, check out Peter's work on 'Wild Food Capture' on Facebook.

1. **Double check that foraging is allowed where you are** — And consider risks you might not be able to see, like spraying.
2. **Go slow, don't rush** — Get a positive ID every time, even for species you've foraged before.
3. **Check and double check every attribute of the fungus** — don't rely on just one or two features to make an identification.
4. **Carefully check each individual fungi you find** — This is especially important if you find a cluster of mushrooms. You don't want a poisonous interloper to end up in your basket.
5. **Check your finds again when you get home, to re-confirm the ID you made in the field** — If you're lucky enough to find several edible species on one mission, store each one in a separate container to make this step easier and more organised.

6. **Don't rush the foraging process, and don't rush into trying your finds, either** — Try a bit, see how it sits. If you like the taste and it agrees with you, have a bit more next time. Even some choice edible species, like porcini, have been known to upset some tummies.
7. **Cook your finds thoroughly before eating** — Many wild fungi species have been known to cause stomach upsets when raw or undercooked.
8. **Be aware of the key poisonous species found in Aotearoa, listed on page 85** — But remember, ruling these out does not mean you have found an edible fungi. And, of course, when in doubt, leave it out.

4

Edible fungi

Skipped straight to this bit? That's understandable. Wild mushrooms are some of the most romantic and special ingredients out there. But this is not an area to rush through. Use the previous sections to get up to scratch on your fungi identification and foraging skills, *then* dive into this section to learn more about the edible fungi of Aotearoa.

As we've learned in this book, fungi can make delicious and nutritious food. I've loved learning how to forage for fungi and remember lots of my finds individually. From field mushrooms to tawaka, there are more than a few tasty species to find out there.

Of course, fungi foraging is not a new practice. Māori culinary tradition includes fungi as food (and medicine) alongside the other bounties the native bush offers. Over a dozen fungi species have been recorded as being used in Aotearoa as food resources.[1]

Going slow while foraging fungi is always key. Double check you're allowed to forage in the area where you're looking. Investigate if the area might be sprayed. Take your time to get a positive ID. And make sure you read pages 88–91 for information about how to forage respectfully, reciprocally and safely.

In my experience, foraging fungi is a game of inches. It's unlikely that you'll find multiple new-to-you edible species that you can positively identify on just one walk. I think of it like making good friends. It's hard to add a bunch to your inner circle all at once. And if you hurry, you could end up with some bad/toxic ones in the lot. I've only ever been able to add one species at a time to the group I 'know'.

Previous page Hakeke woodear. **Left** Tawaka and other ingredients ready to be made into a mushroom pasta.

And sometimes these additions are spaced out by months or even entire seasons. Experimenting with foraged fungi in the kitchen is a whole other adventure. I'm still learning about fungi to forage, and how to use them. It's a foodie journey without end, really.

Fungi as food

The world is your oyster mushroom. That's how one top New Zealand chef described the world of edible fungi to me. Heaps of us have unhappy memories of slimy shrooms being served up to us as children. Even me. Fungi, though, make fantastic and often surprising food.

Here are some of the most interesting fungi foods I've come across in Aotearoa. Slippery jack mushroom burgers, grilled over charcoal, with a dash of pine oil, served over a bed of creamy mushroom-stock polenta. Mushroom mince dumplings. A porcini mushroom chocolate mousse Yule log. Those first two dishes come from Max Gordy, and the third from Vicki Young — both are top Wellington chefs. When we think outside of the 'mushrooms on toast' box, we find that fungi offer us untapped foodie potential.

People who don't like mushrooms just might not have tried the right one yet. There's such a wide scope out there, a huge amount of diversity, and they're good for you too. To cook with mushrooms at home and make them sing, follow these tips gathered from some of the best chefs and fungi fans around Aotearoa.

- Tidy them up — use a tea towel or pastry brush to remove dirt or debris. If you want to wash them, wash them gill side down, otherwise water will get trapped in the gills and make them slimy.
- If you want your mushrooms to remain firm, sweat them off. Put them in a hot pan with a bit of oil and cook them out, but don't add salt — the salt will draw out the moisture and the mushrooms will quickly go soft and stick to the pan.
- Dice them up to make a mushroom mince. Add a bit of oil to your pan to spread the heat around evenly. When

it's hot, add your mushrooms. Once you start to see them caramelising on the edges, add a bit of water, white wine or vinegar to de-glaze the pan. You can add your mushroom mince to just about anything.

- If you've got heaps of mushrooms, you can always dehydrate them. Slice them, then leave them in a sunny corner on a baking tray, or in the oven with the door open at a low temp and under supervision. Store and add to soups and stocks. Or powder them to make your very own umami sprinkle.

Growing your own

The journey of finding, identifying and safely foraging edible fungi is fun, but it isn't fast. It takes a while to build up your knowledge and skill. In the meantime, though, you can grow your own edible mushrooms. Little mushroom farms doing just that can be found up and down Aotearoa. When I first moved to Ōtautahi Christchurch, I didn't really know anyone. So when the owners of a small mushroom farm offered to show me their operation, I went.

Taylor and Susan, who are now my friends, started Sporeshift Mushrooms to grow fungi food in a way that's good for people and the planet. They taught me that growing mushrooms can be super-sustainable. You can grow indoors in a controlled environment with little waste, and you can use vertical space, too, which massively increases output per hectare. Agricultural and forestry waste products can be used as the growing substrate, and what remains after the mushroom harvest can be turned into nutritious garden compost.

Almost anyone can get into mushroom-growing. Starting out with a mushroom grow block is probably the way to go — these can be purchased online from Sporeshift and other producers around the country. If you go well with those and really want to go further, you can make your own grow blocks. This is a big jump up, though — you'll need a flow hood, a sterile environment and lab equipment.

If you're not keen to turn your garage into a lab, try co-planting in the garden. Studies have shown that wine cap mushrooms, for example, can boost corn's ability to produce delicious ears. These mushrooms are prolific decomposers, they digest organic matter in the soil, and this adds nutrients to the system that the hungry corn eagerly convert into sweet kernels. To start companion planting with fungi, add woodchips and mushroom spawn to your soil. This can be purchased online. Wine caps, and other tasty species like the phoenix oyster mushroom, can be grown this way.

How to use this section

There are heaps of delicious fungi to forage for in Aotearoa. The ones featured in this book are some of the most well known. Use this guide to learn more about them, and maybe even to try one! But not before you get a positive ID, of course.

For each species in this section, I've included as many of their names as possible. Some of our native fungi have recorded Māori names. I've included these, but want to note that te reo Māori has many dialects so not all names may be listed here. Some species also have quite a few common names; I've listed the most popular.

The species featured here are grouped by the environment you're likely to find them in: native bush, fields and pastures, or urban/developed areas. Within this, each species is ranked as easy, medium or hard to forage. This rating factors in how common the species is, how simple or tricky it is to identify and whether it has any poisonous look-alikes.

The information on distribution and peak season comes from iNaturalist, and will give you a rough ballpark of where and when to look for these beauties. Of course, fungi can't really be put in a box like this — they will emerge whenever the conditions suit them. A warm spell in May, or a chilly band of rain in December — they don't care about the date! Some super-keen fungi-hunters will stick thermometers in the soil, watch the weather patterns, and try to pinpoint the perfect day. But sort of like shares on the stock market, perfect days are near-impossible to pick; the market always wins. Go early, go often — that's my motto.

In these species descriptions you'll find identification information along with a few of these species' stories. These stories have been gathered from previous publications (listed at the back of this book), kōrero and my own observations.

I've also included information on how to use these species. How to harvest, prepare, preserve and enjoy them. When it comes to harvesting, be gentle. Pluck gently or cut; the research shows that both are fine and don't harm the mycelium. And remember, as we've learned in this book, a mushroom is the fruiting body of a mycelial network. Removing the fruiting body won't kill the mycelium. Just like picking an apple won't kill the tree.

Of course, *never* eat anything that you haven't positively identified as being edible. There are deadly poisonous mushrooms in Aotearoa. Proceed with curiosity, care and caution in equal measure. Do some research to find out if the area you're exploring is one where you can forage, and one where it's safe to forage. Think about what you can offer the land you're on, not just about what it can offer you. Hone your identification skills and foraging practice. *When in doubt, leave it out*. Focus on the flow of foraging, rather than just trying to get a feed, and you'll always have fun.

Pekepeke-kiore.

Hakeke

Scientific name
Auricularia cornea

Other common names
Hākekakeka, wood ear, jelly ears, tree ears

Typical environment
Native bush

Difficulty
Easy

Distribution
Found across Aotearoa

Peak season
Always around

Size
Each fruiting body can grow up to 15–20 cm across

In the 1880s this fungus was one of the biggest exports coming out of Aotearoa. It was valuable enough to turn Taranaki farmers into foragers, if only temporarily, and the industry was initiated by just one man: Chew Chong.

Chong, originally from China, immigrated to New Zealand in the late 1800s. Upon moving to Taranaki, he noticed that hakeke wood ear fungus, a prized ingredient in his homeland, was growing from the trees there. It had little to no value in Aotearoa, but if he could get it back to China . . .

From 1870 to 1890, hakeke was collected and sold to China, where it was used in soups and as treatments for a whole host of bodily ailments. The trade Chong set up became so lucrative that hakeke became known as 'Taranaki wool'. From 1872 to 1883 almost 2000 tonnes of foraged, dried hakeke was exported to China. The hours required to gather that much would've been huge — especially when you consider that hakeke loses almost 90% of its volume when dried![2]

Early Māori reportedly used hakeke as a food resource in lean times. It was steam-cooked with pūha, rangiora and other greens, and was sometimes given to treat tutu and karaka poisoning.[3]

How to find and identify

Environment: Look in the bush or anywhere with plenty of dead or decaying wood. This fungus often grows from tawa, pukatea and māhoe trees.

Top: Soft, rubbery, jiggly, and brown when there's been rain; looks almost like an ear. Shrivelled, crusty, darker and smaller when there's been no rain or if it's an older specimen. Sometimes has super-fine downy hairs on it.

Spore surface: The spores are held beneath the cap. The underside of the cap is smoother and lighter than the top, and the flesh is semi-gelatinous.

Spores: Dusty white, can sometimes be seen on the underside.

Stipe: None.

How to use

Hakeke shines when paired with the right flavours. It adds crunch and can be turned into a tasty salad when sliced thinly and served with sesame oil, chilli and coriander. Dishes like this can be found on some Chinese restaurant menus and this species is sometimes listed as 'black fungus'. Hakeke is easy to dehydrate in the sun and can be rehydrated in ramen, stock, soups and curries.

Pekepeke-kiore

Scientific name
Hericium novae-zealandiae

Other common names
Coral tooth fungi, fungus icicles, monkey head, hedgehog fungus

Typical environment
Native bush

Difficulty
Easy

Distribution
Found across Aotearoa

Peak season
March to July

Size
Can vary a lot, from about 5 cm to 25 cm across

Pekepeke-kiore grows in dense bush. Its white, sparkly sheen stands out against the green flora. To me it looks like a fungal waterfall, frozen in time, or a miniature castle for fairy ice-queens.

The Māori name gives this fungi another persona. 'Pekepeke' means to jump around, or hop like a little bird, and 'kiore' refers to a rat. Together the words call to my mind a furry critter, darting across the forest floor; suddenly startled, he jumps into the air, his fur pointing every which way.

Pekepeke-kiore is as tasty as it is visually striking. There are a few reports of Māori using it as a food resource, and it may have medicinal applications, too. Pekepeke-kiore is closely related to lion's mane, *Hericium erinaceus*, which has been extensively studied for its potential to support brain and nerve health. It has been shown to boost cognitive functioning and help regenerate damaged nerves.[4]

How to find and identify

Environment: Found in native bush, introduced woodlands and old growth pine forests. Remind yourself to look up on bush walks, and you might spot one! Scan the forest floor, too. If you spot what looks like a soggy cauliflower, you might be getting closer. These fungi often fall to the ground in a mushy heap as they age.

Body: This fungus doesn't have a cap and stipe as such. Its form and intricate spines are a dead giveaway. Colour ranges from white to buff.

How to use

Pekepeke-kiore is tasty when young and can be pulled apart like pork. It's chewy, tender, juicy and tastes a bit like cauliflower. You can deep-fry bits of it and serve as fritters with aioli. Or shred it and make vegan 'crab' cakes. Whatever way you cook it, it's hard to go wrong with a bit of crumb, hot oil, and a good dipping sauce.

Tawaka

Scientific name
Cyclocybe parasitica

Other common names
Poplar mushroom, sword belt mushroom

Typical environment
Native bush

Difficulty
Easy

Distribution
Found across the NI and in Nelson, Tasman, Marlborough, Canterbury and Otago in the SI

Peak season
November to May

Size
Cap: approximately 8–12 cm across

Tawaka mushrooms are dinner-plate-sized delights. They're fairly easy to identify and can be found growing from both native and non-native trees, which means there are heaps of opportunities to spot them. A study at Lincoln University found tawaka dry mass to be 20% protein.

Early Māori may have gathered tawaka in summer and steam-cooked them. If someone ate tawaka and walked through the māra kai (garden), it was said that all the gourds would shrivel up on their runner vines. None of the kamokamo in my garden grew the summer I first tried this fungi.

In addition to being delicious, tawaka may also have medicinal applications. It was reportedly given to treat fevers and to fortify expecting mums. It was also used to combat karaka and tutu poisoning.[5] Tawaka is a great grow-at-home mushroom; kits are available online to help you get started.

How to find and identify

Environment: If you're in the North Island, look for tawaka growing in clusters on native hardwoods like tawa and introduced trees like willow and poplar, which are often planted along fence lines. In the South Island, this species usually grows in coastal/lowland environments, often on lacebark trees. Don't forget to look up, too! This species often grows at eye level or above reach (devastating).

Cap: Very large, smooth and almost cool to the touch. Domed and darker brown when young; flatter when fully opened, with a light brown-buff colour.

Spore surface: Has light-coloured gills. On younger specimens these are covered by a veil. The gills are adnate.

Spores: Dark, chocolatey brown.

Stipe: Thick, woody; has a creamy white colour and usually has a sizable skirt that features a coating of dark brown spores. This dramatic skirt is one of the most distinctive traits of tawaka, but it sometimes falls away over time.

How to use

Tawaka is a top culinary mushroom. They're large enough to make a mean mushroom burger — simply brush with marinade and grill whole.

Tawaka brown butter pasta is also pretty special. Melt a knob of butter and heat until it takes on a toasty colour and a nutty aroma. Add salt and a handful of fresh parsley, basil or sage to it, and set aside. Tear your tawaka into strips, put a bit of brown butter in a hot frying pan, then add the tawaka and sizzle until the edges are crispy. Mix these tasty morsels and the remaining browned butter through nice cooked pasta, and top with hard cheese.

Brown oyster

Scientific name
Pleurotus australis

Typical environment
Native bush

Difficulty
Medium

Distribution
Mostly in the Northland, Auckland, Waikato, Manawatū and Wellington regions. There have also been a few sightings in Nelson and Tasman

Peak season
October to July

Size
Cap: up to 20 cm across

This native fungi is a choice edible. Many people who typically don't enjoy mushrooms find they actually like *Pleurotus* species, which are a lot less slimy. These species get their common name from their delicate flavour, which has slight seafood notes to it. The fruiting bodies are also shaped a bit like oyster shells from above.

The brown oyster often grows from mānuka wood in native bush. They sometimes grow in tight clusters that are stunning to photograph. If you're impatient to try these, head to your local farmers' market. Small-scale mushroom farmers often sell oysters, and some even sell grow-your-own kits. These kits are heaps of fun as the mushrooms change almost daily, going from tiny pinheads to full-sized fruiting bodies in just a few days.

How to find and identify

Environment: Look for this species growing from dead wood, particularly mānuka, in native ngāhere.

Cap: Dry and smooth to the touch, with a tawny brown colour. The edge of the cap is often noticeably rolled 'in' towards the gills when young, but this becomes less noticeable as the cap stretches to open more fully. It's fleshy and shaped like a shell.

Spore surface: Has deeply decurrent gills which are cream to brown in colour.

Spores: White, but are more pale tan when dry.

Stipe: Doesn't have a stipe per se but has a short stubby stem that connects it to the growing medium. The stem often has faint gill ridges on it. The cap is slightly offset from the stem meaning the stem connects to the cap not in the direct centre, but towards the back, closer to the growing medium.

How to use

Oyster mushrooms, including this species and varieties that can be grown at home or purchased at farmers' markets, are great for making 'bacon'.

To make this vegan treat, tear each oyster into strips — the rough edges will add surface area and hold more flavour. Heat butter in a pan until it's very hot, then add the mushroom strips. Cook until the edges are crispy, golden and crunchy. Add to pastas, sammies, omelettes, whatever bacon goes well with, really!

Velvet shank

Scientific name
Flammulina velutipes group

Typical environment
Native bush

Difficulty
Hard

Distribution
Found across Aotearoa apart from Northland

Peak season
April to September

Size
Cap: between 3–9 cm across

Velvet shank can freeze solid and still survive. It's one of the best winter fungi you can find, and you may already know its famous cousin enoki. Enoki is elegant. Each extremely slender stalk is topped with an ultra-tiny, snow-white cap. They grow in bunches and are often used in tom yum soup and many other stews, broths and hot-pots. Enoki has been cultivated in China for over 1000 years.

Velvet shank and enoki are closely related, and yet, they look so different. This might be a result of their different growing environments. Enoki is grown commercially, at scale, and inside. Velvet shank, meanwhile, grows outside, in ever-changing conditions, exposed to wintry winds, rain and snow. These different environments draw out different expressions from blueprints that almost match. So cool.

If you're foraging for velvet shanks, just be aware that there is a look-alike species called the funeral bell (*Galerina marginata*, see page 85). Obviously not one to mess with!

How to find and identify

Environment: Keep an eye out in damp, dense bush. This fungus tends to grow from dead or decaying hardwood trees. They grow in cute clusters, not alone.

Cap: Yellow to orange brown, often darker in the centre. Has a slimy, slightly sticky texture; sometimes bits of dirt and tiny bugs get stuck to it.

Spore surface: White gills that turn creamy yellow with age. Some run all the way from the edge of the cap to the stipe, others don't make it all the way there. The gills have an adnexed to sinuate attachment to the stipe.

Spores: White.

Stipe: Light yellow at the start, turning dark brown to near black with a velvety texture over time. Quite tough. No skirt.

Dangerous look-alikes: Funeral bell (*Galerina marginata*, see page 85), could be confused with velvet shank. Funeral bells have brown spores and skirts, but velvet shanks have white spores and no skirts.

How to use

Velvet shanks offer delicious texture and flavour. They are great in tom yum soup, or any brothy, hot-pot-style dish you might be making. For a quick side dish, tear them into smaller pieces. Heat sesame oil in a pan, add your mushrooms along with a dash of soy sauce, garlic and a tiny bit of ginger. Top with a few sesame seeds, chopped spring onion and fresh chilli. Serve over rice.

Tawaka cap, skirt and spore print.

Tawaka.

Viscid (sticky) caps of velvet shanks.

Giant puffballs

Scientific name
Calvatia gigantea

Other common names
Pukurau, Poketara

Typical environment
Fields and grassy edges along paths, gardens and glades

Difficulty
Easy

Distribution
Found across Aotearoa

Peak season
January to June

Size
Can grow to 50 cm across, or larger in some cases

Giant puffballs might be Mother Nature playing a prank on us. They are straight-up hilarious. The largest one ever documented in Aotearoa was a whopping 7 kg. And they're one of the most sexually active organisms on earth. Each one contains trillions of reproductive spores.

Giant puffballs are also known as 'tofu of the woods' thanks to their spongy texture, nutty flavour and relatively high nutrition. They may have been used by Māori as food, and to stem bleeding and treat scalds and burns. Pukurau and poketara are two Māori names for fungi in the *Calvatia* genus. Globally, there are many reports of puffballs being used as a styptic. The dried powdery spores were used in early beekeeping, too, as their anaesthetic quality would subdue the bees during honey harvesting.

The Hawke's Bay town, Waipukurau, may even get its name from these oddball funguses. Early Māori reported that pukurau grew along the river there and were washed in the water to soften the flesh before cooking.[6]

There are several *Calvatia* species in Aotearoa, including the giant puffball and two slightly smaller native species.

How to find and identify

Environment: Look in fields, paddocks and lawns for these soccer-ball imposters. They sometimes hide or nestle in tall grass.

Body: Huge, globular, white, smooth. Does not have a classic cap or stipe; is simply a fungal orb. There is a connection point with the ground that might have a gnarled texture and a thready root-like filament coming from it, but it's not a stipe per se.

Flesh: Smooth, semi-crumbly, and completely white all the way through. If the interior flesh has taken on a yellow, green or brown colour, this indicates that the spores have formed and it's no longer good for eating.

Dangerous look-alikes: Smaller specimens could be confused with *Amanita* species that have not yet emerged from their volvas. Slice them open to check: if you see the faint outline of a mushroom, you do *not* have a puffball. Earthballs, which are poisonous, are sometimes also confused with puffballs, but earthballs have a brown, warty exterior.

How to use

Giant puffball parmesan sammies are hard to beat. Slice your puffball thinly, dredge each slice in panko breadcrumbs, then shallow-fry. Put these on some sourdough along with marinara sauce, mozzarella, fresh herbs and parmesan. Giant puffballs also do beautifully on the barbecue. Cut them into steaks, brush with a glaze of miso, soy and honey, and grill on each side for 1 to 2 minutes. The easiest option is to cube your puffball and sub it for tofu in any curry, laksa or soup. Puffballs don't get the culinary limelight very often, but in the right company they truly shine.

The author with giant puffballs – the largest ones were too old to eat.

Grilled and glazed giant puffball steaks.

Shaggy inkcaps past eating stage,
but perfect for ink-making.

Shaggy inkcap

Scientific name
Coprinus comatus

Other common names
Shaggy mane, lawyer's wig

Typical environment
Fields, pastures, the gravelly edges of tracks, areas where the soil has been disturbed

Difficulty
Easy

Distribution
Found across Aotearoa

Peak season
February to June

Size
Cap: 4–8 cm across and 6–20 cm tall; Stipe: between 6–40 cm long

This fungus had a shag hair-do way before the 1980s. Its layered mane is iconic; but just like the hair trend, this mushroom doesn't last long.

Go for a fungi hunt in the morning and you might spot some shaggy inkcaps poking through the dew. Their fringy white caps and impressive height help them stand out against green grass. Return in the afternoon, though, and you'll find they've turned into inky black messes. They go from groovy to goth in just one day.

This process is called auto-digestion or deliquescing. First an inkcap emerges, then it opens, starts to drop its spores, and then it begins to dissolve itself from the bottom of the cap upwards, creating a curling effect. This helps the final spores catch the wind as the rest of the mushroom melts into a puddle of goo like the Wicked Witch of the West.

Shaggy inkcaps don't tend to have a lot of fame in the world of foraged fungi because of the messy ink issue, but I've found a few fun ways to use them. I actually rate their flavour, too. It's delicate, and pleasantly mushroomy without being overwhelming.

How to find and identify

Environment: Look down! Shaggy inkcaps grow from the ground, usually in open, grassy areas. They frequent urban environments like parks, as well as grassy fields.

Cap: Has a unique shaggy texture which will come off in your hands. When these first emerge, the cap is off-white and clings tightly to the stipe at the bottom, in a torpedo shape. Within a few hours the cap opens away from the stipe, and begins to turn black and curl upwards from the bottom.

Spore surface: Gilled, white to light pink when young, turning inky black with age. The gills are free of the stipe — cut the specimen lengthwise to check this feature.

Spores: Blackish brown.

Stipe: White, slender, sometimes hollow, sometimes slightly swollen at the base, and often quite tall. Has a small ring-shaped skirt which doesn't last long.

Look-alikes: The common inkcap (see page 201) has a smooth cap, with no shag, and should not be consumed with alcohol.

How to use

Fry firm specimens in butter and eat atop toast. If you're not going to use them straight away, remove the stipes and pop them in the fridge; this will extend their shelf life by a day or two at best.

Use older, slightly inky ones to make mushroom stock. Melt a blob of butter in a pan. Add crushed garlic, chopped onion and red pepper flakes, and simmer until it smells nice. Add your shaggy caps plus a cup of milk or cream. Simmer to reduce. Use this to make a navy-coloured risotto, or serve over steak with plenty of cracked pepper. If you're extra-fancy, you could use the ink as vegan squid ink. Or add water to it and paint with it!

Fairy ring champignon

Scientific name
Marasmius oreades

Other common name
Scotch bonnets

Typical environment
Fields and pastures

Difficulty
Medium

Distribution
Found across Aotearoa

Peak season
November to May

Size
Cap: 1–5 cm across; stands about 8 cm tall on average

Chefs go nuts for fairy ring champignons. They have a delicate almondy flavour and are oh so pretty. They're semi-easy to find, and lightweight, which means they can be dehydrated in the sun in an hour or two. Fairy ring champignons always grow in a community, never alone. So if you find one you find many, and that's handy when you're foraging.

These graceful mushrooms don't just grow in haphazard bunches — they grow in circular 'fairy ring' patterns. Fairy rings usually occur in fields where there aren't many roots or rocks in the soil below. In that setting, a mycelial network is free to expand equally in all directions from its central point. Then, when the temperature and moisture levels are right, the network sends up fruiting bodies (mushrooms) at its outermost edges. And that's what we see above ground — a mycelial moment frozen in time, a window of opportunity, a ring of delicious mushrooms.

Historical explanations for fairy rings are quite funny. Some people thought they were tendrils of subterranean vapour, dance floors for witches, goblins and elves, or the work of slugs, ants and moles.[7] This species is only to be eaten in moderation, though, as it can sometimes contain low-level toxins.

How to find and identify

Environment: Look in open grassy areas, anywhere with uncrowded yet rich soils. You won't find these in the deep, dark woods or growing alone. Be sure to look out for a circular, ring-shaped growth pattern.

Cap: Pale tan colour, conical in shape; sometimes has a raised bump in the centre, called an 'umbo', that is darker than the rest of the cap.

Spore surface: Has white to tan gills. The gills do not touch the stipe at the centre, though; they are 'free' from it, and this is a key identification feature.

Spores: White.

Stipe: White to off-white; thin, yet fibrous and tough — it's tricky to separate from the cap without tearing it. No skirt.

Dangerous look -likes: There are heaps of little brown mushrooms (LBMs) out there that could be confused with these. Be sure to check for a circular growth pattern, an umbo, free gills, a fibrous stipe and so on, to confirm that you've got fairy ring champignons.

How to use

Use fairy ring champignon's nutty-tasting notes to put a twist on green beans almondine. Steam a handful of green beans, gently fry your mushrooms with butter, combine, and season to taste. You can also toss them through angel hair pasta with fresh tomatoes, olive oil and parmesan for a special summertime-inspired pasta. You could even use these as garnish for desserts.

Field mushroom

Scientific name
Agaricus campestris group

Other common names
Fieldies, paddock mushrooms

Typical environment
Fields and pastures

Difficulty
Medium

Distribution
Found across Aotearoa

Peak season
March to July

Size
Cap: 3–12 cm in diameter; Stipe: about 3–10 cm long

Farmers and country folk have been foraging fieldies for generations; they're part of New Zealand's pastoral tradition. Ask anyone how they identify field mushrooms, though, and you'll always get the same answer: 'I just know.'

This response freaked me out the first time I heard it. The genus *Agaricus* includes both tasty mushrooms and gnarly, not-for-eating ones. How could anyone 'just know'? But this response is actually a great example of knowledge that's been handed down and become tacit — understood without being stated. Lots of Kiwis learned to forage fieldies as kids; they simply have the feel for them.

We don't know exactly what *Agaricus* species we have/don't have here, so a microscopic look is required to determine the particular species. When it comes to fieldies, though, it's all about knowing what to check for, what to watch out for, and taking your time to get to know them. Even if you do 'just know them', you still need to check and get a positive ID; death cap mushrooms have been mistaken for fieldies before. Yikes.

How to find and identify

Environment: Look in grassy, open fields.

Cap: Fleshy, with a white to buff-brown colour. It sometimes has a downy texture which hangs over the edge a bit.

Spore surface: Has gills that range from light pink when young to dark brown, near black, with age. The gills are free of the stipe.

Spores: Chocolatey brown.

Stipe: Emerges directly from the ground — check to make sure it is *not* growing from a volva/egg sac, as that could indicate you have a death cap. There should be a skirt, or evidence of one, although sometimes it's faint.

Stain: Bruise the cap and the base, which should stain light pink. Check to make sure it does *not* turn bright yellow, as that could indicate you have a poisonous yellow-stainer.

Scent: If the mushroom smells acrid or like an indoor pool, toss it out — it might be a yellow-stainer. Some people can't smell this, though, so don't rely on this feature alone.

Dangerous look-alikes: The death cap could and has been confused with field mushrooms. To rule out the death cap, ensure that your mushroom is not emerging from a volva and does not have white gills. The yellow-stainer could also be confused with field mushrooms — to rule this one out, ensure that your find does not stain chrome yellow and does not have a chlorine-like smell.

How to use

Field mushrooms are easy-as to cook with. Add to pizza, pasta or stir-fries. Or toss them in a pan with butter, rosemary, plenty of salt and pepper, and serve them up on thick toast. This Kiwi café classic simply cannot be beaten.

Horse mushroom

Scientific name
Agaricus arvensis group

Typical environment
Fields and pastures

Difficulty
Medium

Distribution
Found across Aotearoa

Peak season
September to May

Size
Cap: 7–20 cm across; Stipe: 5–12 cm long

The horse mushroom is the beefcake cousin of the field mushroom. It's meaty and delicious, and tends to pop up in the same environments as fieldies — open, grassy expanses.

This species gets its common name from where it likes to grow: paddocks and fields that have rich soils and are well fertilised. These mushrooms, like horses, are also HUGE. Just one can make a meal, and when you spot one, you're liable to find another — they often emerge in fairy rings.

The most distinctive feature of this mushroom is the partial veil, which covers the gills initially. As it begins to break away from the cap, the veil looks like a cog wheel or toothy gear. Once fully broken from the cap, the partial veil falls and its remnants skirt the stipe.

Genus *Agaricus* contains some of the best edible fungi, like the horse mushroom and the field mushroom, but it also contains poisonous species. And it's important to remember that we aren't entirely sure exactly which *Agaricus* species are or are not present in Aotearoa; there's an estimated 60 species. To gain certainty at the species level, you'd need to use a microscope. For field identification of this mushroom, though, I check for the features I expect it to have and rule out dangerous species that could be confused with it. These are mainly the yellow-stainer and the death cap.

How to find and identify

Environment: Look in lush paddocks and open grassy expanses. This species likes nutrient-rich soils.

Cap: White to greyish brown, can be smooth or a bit scaly; starts out very globular and opens flatter with age.

Spore surface: Gills that start off near-white but quickly turn pinky grey, then brown. The gills are free of the stipe. Initially they are covered by a partial veil.

Spores: A dark purple-brown colour.

Stipe: Stout, portly, light in colour. Should have a large, thick skirt, but these sometimes shrink or fall away over time.

Stain: The interior flesh should be white but may stain slightly yellow.

Dangerous look-alikes: Yellow-stainers and death caps. To rule out a yellow-stainer, bruise the cap and base — if it stains bright yellow, or smells like chlorine, chuck it out. Death caps have white gills, white spores and emerge from a volva, while horse mushrooms have brown gills, brown spores and do not emerge from a volva.

How to use

Horse mushrooms can be used just like field mushrooms. Top a pizza with them, add to stir-fries, make a cream of mushroom soup, or stuff these large caps with rice, mince, chopped capsicum and spices, top with cheese and bake in the oven.

Fairy ring champignon gills and umbo.

Field mushroom caps can sometimes be downy.

Free gill attachment of fairy ring champignon.

Birch bolete

Scientific name
Leccinum scabrum

Typical environment
Urban/developed areas

Difficulty
Easy

Distribution
Found across Aotearoa apart from Northland

Peak season
February to May

Size
Cap: 5–11 cm across

The birch bolete deserves a new common name: old reliable. Just when you think your luck has run out and there are no edible fungi to find, a birch bolete appears. Not the most delicious or highly sought-after shroom, but always/often there, and occasionally huge, too!

Beginner-friendly, this mushroom is easy to spot, identify and cook with. It's a fun, no-frills find, and if you know how to identify a silver birch tree, you're already halfway there. Their shimmery, papery bark will guide you straight to these edible mushrooms.

Once you've found a birch bolete patch, visit again. They flush reliably and can be used as an indicator that it's a good time to go for a bigger hunt. When birch boletes emerge, other species like porcini tend to follow.

How to find and identify

Environment: This species specifically grows with silver birch trees. So check their bases.

Cap: Smooth, firm, chestnut-brown when young; gets flatter, squishier and slimy with age or after rain. These caps tend to blend in with autumn leaf litter, so give yourself a few minutes to really get your eyes on when out looking for these.

Spore surface: Has pores rather than gills; these are off-white and very small when young, but become enlarged and yellow-brown with age. If you find an older specimen that has started getting squishy, you can easily peel the pore surface away from the cap. Chuck the pores away and work with the firmer flesh that remains.

Spores: Brown.

Stipe: Can be quite bulbous at the base when young, getting straighter and more barrel-shaped over time. The flesh is white, and features a distinct black speckling that has a rough texture. This characteristic will help you identify this species against other boletes.

Stain: May not stain at all when bruised, or may turn light pink.

How to use

This species is sometimes seen as porcini's ugly cousin, but I think they're nice in the right company. If you get a firm specimen, slice it thinly, make a red wine jus, add your mushrooms, cook thoroughly and pour over a nice steak. Or get your pan nice and hot, dry-fry the mushroom slices until they start to look like bacon, then add to pizza. If you find a squishier specimen, dehydrate it, powder it, and sprinkle it over soups and risottos for an umami kick.

Slippery jacks

Scientific name
Suillus granulatus and *Suillus luteus*

Other common names
Dotted-stalk suillus and purple-veiled slippery jack

Typical environment
Urban/developed areas

Difficulty
Easy

Distribution
Found across Aotearoa

Peak season
January to July

Size
Cap: 4–12 cm across; Stipe: 4–8 cm long

Find a slippery jack spot, and you'll never need another. These edible species love company and they often grow in scores. While they're not the most prized of wild mushrooms, their colour is lovely and they're great for beginner foragers.

Slippery jacks and other edible boletes, like the birch bolete, aren't particularly prized here. They're easily overshadowed by porcini. But if you know how to work with them, you can make something special. Eastern European culinary traditions have included these mushrooms for ages, and appreciate them for their ability to bulk out hearty winter dishes and even to take the place of meat.

Slippery jacks can serve as a great base species if you want to try making your very own umami mushroom 'spice' blend. It's easy to harvest quite a few of these at once, and their odd texture makes them a great candidate for dehydration and powdering. You can then add other mushrooms like porcini or fieldies to the mix to punch up the flavour profile. There are four *Suillus* species that grow in association with pine trees in Aotearoa, but the two mentioned here are the most common.

How to find and identify

Environment: Look beneath pines and in pine plantations, but check that the area hasn't been sprayed.

Cap: Slimy, slippery and dark brown; flattens out with age.

Spore surface: Has pores. The surface has a distinct yellow colour, and the inner flesh has a bright, creamy yellow colour.

Spores: Brown.

Stipe: The dotted-stalk suillus has unique tiny brown granules towards the top of its stipe, and no skirt. The purple-veiled slippery jack has a partial veil that initially covers the pores, then falls away and has a unique purple-brown colour.

Stain: Bruise the spore surface — check that it does *not* stain blue. Some inedible boletes stain blue.

How to use

To use slippery jacks, peel off the slimy cap — some people have noted that this bit can cause indigestion and it generally has an odd texture anyway. You can also peel the squishy pores away. You'll be left with the inner flesh, which tends to be firmer and nicer to work with. Think of this species less as a main act, and more as a supporting ingredient that can add depth and volume. Add them to mushroom stocks, ramens or curries.

Fleur Sullivan, founder of Fleurs Place in Moeraki, picked these often as a child in Central Otago. She reckons the best thing to do with them is to dry and powder them, or smoke them and add them to risotto.

Suillus granulatus.

Peppery bolete

Scientific name
Chalciporus piperatus

Typical environment
Urban/developed areas

Difficulty
Easy

Distribution
Found across Aotearoa apart from in the Northland and Gisborne regions of the NI

Peak season
January to June

Size
About 7 cm x 8 cm

If you like to go beyond Kiwi-hot, then this might be a mushroom for you. The peppery bolete brings spicy goodness to the world of edible fungi.

This introduced species can be found in the warmer months and can be used as a spice or condiment, but should not be consumed in large quantities. It may actually be a parasite of the fly agaric (*Amanita muscaria*, see page 210). Either way, the two species seem to like very similar environments, so let the highly visible red and white toadstools light the path towards these peppery specimens. Do be aware, though, there are some reports of this species causing gastric upset. It should always be cooked thoroughly and should only be used in small quantities.

This fungi has also been used to make dye. Depending on what mordant is used, yellow, orange and green dyes can be obtained using the peppery bolete.

How to find and identify

Environment: These grow in groups or alone, on the ground, beneath introduced conifers, beech, birch and oak trees.

Cap: Has a cinnamon red to brown colour. Has a unique sheen to it, even when it's dry. Sometimes cracks with time. The interior flesh is bright, chrome yellow from the stipe upwards, but becomes more red towards the top.

Spore surface: Has cinnamon to brown coloured pores.

Spores: Brown.

Stipe: Bright yellow at the base, but this is sometimes hidden by the earth and leafy debris. The rest of the stipe is yellow to rusty cinnamon with specks of red. The stipe has a vertically striated look to it.

How to use

This mushroom isn't a main-event species. Its peppery profile though can be used in soups, stews and other savoury dishes. If you're making a creamy pasta sauce, for example, you could add a few of these to give it a bolder bite. Or dice it and add it to veggies to elevate their flavours in a stir-fry. Cooking does diminish the peppery flavour of these boletes, however it's important to cook this species thoroughly and only consume a bit as it has been known to cause gastric upset in some people. Think of this species like a condiment. Build other, bigger flavours with its help. Or even season other mushrooms with it.

Porcini

Scientific name
Boletus edulis

Other common names
Ceps, penny buns

Typical environment
Urban/developed areas

Difficulty
Easy

Distribution
Wellington in the NI, Canterbury and around Dunedin in the SI

Peak season
February to June, and November to December

Size
Cap: 7–30 cm across; Stipe: 8–25 cm long and up to 7 cm thick

Nothing gets foragers more fired-up than porcini season. Porcini means 'little fat pig' in Italian, and they're some of the most famous culinary mushrooms out there. Their deliciousness has been starring in European dishes for centuries. So, how'd they get *here*?

When European colonisers were establishing Christchurch, they wanted the city to remind them of home. Hagley Park and its botanical gardens were planned and installed to reflect the mother country. Mature trees were harvested and shipped over from Europe to complete the vision. And in at least one tree's hessian-covered root ball was a stowaway — porcini mycelium.

Today, porcini can be found around some of our main urban centres. They emerge in autumn and bring with them a flush of excited foragers, who are often quite competitive and may arrive at first light — or even before, with the help of headlamps — in the hope of finding a few wee porcini. Luckily, these tasty mushrooms can also be purchased in dehydrated format from speciality grocery stores.

Porcini can be hard to see since they blend in with the leaf litter, but here's a top trick: they like similar environments to *Amanita muscaria* (the big red and white toadstools on page 210). Use these bright caps as an indicator species . . . beacons to guide your hunt.

How to find and identify

Environment: Porcini grow on the ground in association with oak, pine, silver birch, sweet chestnut, plane and European beech trees. They're usually found in areas with mature introduced trees (think fancy old residential areas and parks).

Cap: Chestnut brown, sometimes white at the edge. Young ones are burger-bun-shaped; older ones are wider and flatter. Sometimes slightly sticky to the touch.

Spore surface: Buff white with very tiny pores; the pores enlarge with age and become more yellowy.

Spores: Greenish brown.

Stipe: The texture on the stout, bulbous stipe is a key identifying feature: it's lacy and delicate-looking, is called 'reticulation', and is usually most visible at the top close to the cap. The interior flesh of both the cap and the stipe is white.

How to use

To cook with porcini, follow the advice of my good friend Alba, who's an Italian chef and handmade pasta artisan. Keep it simple, she says — let the mushroom sing! Use fat to carry its flavour far and wide. And remember, you only need a little porcini to make a dish amazing.

Alba's porcini pasta is a no-brainer. Simply brown some butter, add some fresh herbs to it in the final minute of cooking, simmer the mushrooms in the butter, mix through delicious pasta, and grate good parmesan on top.

Don't despair if you find a porcini that's a little past it. Remove the squishy bits, then slice the rest and dehydrate to use later.

Wine cap

Scientific name
Stropharia rugosoannulata

Other common names
Garden giant, king stropharia, Godzilla mushroom

Typical environment
Urban/developed areas

Difficulty
Medium

Distribution
Found across the NI, and in Nelson, Christchurch and Dunedin in the SI

Peak season
March to June, and September to December

Size
Cap: up to 30 cm across; stands up to 20 cm high

Wine caps are winners. They tick every box there is when it comes to foraged mushrooms — they're huge, fairly easy to identify, can be grown at home, and they taste amazing. Wine caps grow among dead wood and have a fondness for mulch beds, which means they're often found in urban environments.

Wine caps aren't just delicious, nutritious food for us, though. They're also pros at breaking down dead wood and turning it into nutrient-rich soil that hungry crops, like corn, require. Several studies have shown that wine caps make great garden companions, and can even support overall forest and ecosystem health. They're one of the best/easiest species you can grow at home, and several Kiwi businesses sell garden-bed grow-kits online.

On the whole, this species is a welcome addition to any forager's basket, dinner plate or home garden.

How to find and identify

Environment: This is a saprobic species, so will be found among decaying organic matter. Look in old piles of wood, and in garden beds that are mulched.

Cap: Has a deep burgundy red colour, akin to an Otago pinot. Starts convex, then flattens out. Can get quite huge — up to 30 x 20 cm!

Spore surface: Grey-coloured gills that have an adnate attachment to the stipe and turn dark purple, near brown, with age.

Spores: Deep purple, near black.

Stipe: Creamy white; will have a skirt, or evidence of one.

Dangerous look-alikes: Wine caps could be confused with *Amanita* species, but the latter have white gills, white spores and often emerge from a volva.

How to use

Antonio Carluccio, godfather of Italian gastronomy and mushroom connoisseur, absolutely loved to cook with wine caps. He referred to them as 'Godzilla mushrooms'. They're large enough to be a meal's main event. Carluccio recommended sautéing these mushrooms with heaps of butter or chucking them on the barbecue. You can sub wine caps for portobellos to make mushroom burgers, or slice them up and add to pasta, sauces, risottos or beef stroganoff.

Wood blewit

Scientific name
Lepista/Clitocybe nuda group

Typical environment
Urban/developed areas with dense leaf litter

Difficulty
Hard

Distribution
Found in the Auckland, Waikato, Manawatū and Wellington regions in the NI; Nelson, Tasman, Marlborough, Canterbury and Otago in the SI

Peak season
April to July

Size
Cap: 4–15 cm across; Stipe: 2–6 cm long

The wood blewit is a particularly special find. For starters, she's purple. How glam! And, like a purple ice queen, she reminds us that there's still plenty of beauty and flavour to be found in winter. Her peak season runs from May to July. If you want to get technical with your hunt, some say that these mushrooms start to come out once temperatures are consistently below 17°C.

Some fungi hunters describe the wood blewit as having a floral, perfumed, fruity smell. The best description I've seen so far is 'sweaty apricot'. I haven't noticed a particular scent myself, but if you find a blewit, give it a sniff.

Despite her dazzling nature, the wood blewit can be found in fairly ordinary environments, like gardens and woody parks. The blewits I've found always seem to be nestling beneath conifer boughs in pillowy pine-needle beds and lofty leaf litter.

How to find and identify

Environment: Grows from the ground beneath both introduced and native trees; loves leaf litter, mulchy beds, pine-needle heaps and the protection of hedgerows.

Cap: Lilac to purple-pink but becoming more tawny brown with age. Convex at the start but opening flat with an umbo (bump) at the top; younger specimens sometimes have an inturned edge.

Spore surface: Has beautiful purple gills, though this hue does fade with age. The gills have a notched attachment to the stipe.

Spores: Off-white to pale pink.

Stipe: Purple-blue with a fibrous texture. Often stout; can be a bit bulbous towards the base. Does not have a skirt.

Dangerous look-alikes: Some *Cortinarius* species could be confused with wood blewit. Take a spore print (see pages 71–73) to rule this out: corts have rusty brown spores, while wood blewits have off-white to pale pink spores. Corts often have remnants of a cortina or veil hanging from the cap like a cobweb, whereas wood blewits do not.

How to use

Wood blewits have a strong mushroomy flavour, and have been known to cause gastric upset in a small number of people, so start with a small amount. And cook them well. This species has a nice firm, meaty texture, so can hold its own as a side dish but can also be added to stews and curries. Cook wood blewits with a squeeze of lemon and eat on toast, or sauté in butter to make a cream sauce.

Shaggy parasols

Scientific name
Chlorophyllum brunneum/ rachodes

Typical environment
Urban/developed areas

Difficulty
Hard

Distribution
Found in the Northland, Waikato, Manawatū and Wellington regions of the NI. And in the Nelson, Canterbury, Otago and Southland regions of the SI

Peak season
December to June

Size
About 15 cm x 15 cm

Shaggy parasols should always be foraged with great caution as there are several dangerous look-alikes. This entry actually refers to two species of shaggy parasol — both are present in Aotearoa. They are indistinguishable from one another without a microscope. Both are edible but need to be foraged with extreme care.

When young, shaggy parasols have a unique concave shape. They have shaggy brown scales on top, are introduced and can often be seen in the warmer months.

These tend to grow in communities, as opposed to on their own, which makes them a handy species for foragers to know about — but it's crucial to distinguish them from the dangerous look-alike species. Additionally, shaggy parasols have been known to cause gastric upset in roughly 1 in 25 people, so try a small bit at first then wait 24 hours to see how it sits. And always cook thoroughly.[8]

How to find and identify

Environment: Look for these growing on leaf litter, often under the protection of tree cover, as opposed to in open fields and pastures.

Cap: When it first emerges, the cap is round and hugs the stipe closely. With time it opens and flattens out and up. The cap's flesh is white with shaggy brown scales on it. Often has a smooth brown umbo, or bump, in the centre.

Spore surface: Gills that are white initially but then turn tan. These bruise a reddish orange, are very crowded together, and are free of the stipe.

Spores: White.

Stipe: Smooth, white to pale brown. Has a small skirt that can be detached and rolled up and down the stipe. The base of the stipe is slightly swollen but it does not emerge from a volva.

Stain: When broken, the flesh rapidly and intensely stains orange/red/brown. This is a key identifying feature.

Dangerous look-alikes: Could be confused with *Macrolepiota clelandii*, but that species does not exhibit the orange/red/brown stain that shaggy parasols do. The vomiter, *Chlorophyllum molybdites*, may be present in NZ and could be confused, but it has green spores. *Amanita* species, like the death cap, could also be confused, as both have white gills and white spores — take extreme care when identifying shaggy parasols.

How to use

To cook with this species when young, remove the stipe and stuff the caps. They're concave, just like a capsicum. Cook some mince with plenty of salt, pepper, onion and garlic. Mix with a bit of rice, a few sliced olives, some chilli flakes. Spoon this into the caps, dot with butter, top with cheese, and bake in the oven for about 20 minutes. This species is known to cause gastric upset in some people, so cook thoroughly, and try just a little at the start.

A small porcini.

A juvenile shaggy parasol.

Porcini slices and fairy ring champignon caps dehydrating in the sun.

Black landscaping morel

Scientific name
Morchella importuna

Typical environment
Urban/developed areas

Difficulty
Hard

Distribution
Not well documented, but has been spotted in Rotorua, Manawatū, Nelson, Christchurch and Dunedin

Peak season
September to December

Size
Cap: 3–15 cm high and 2–9 cm wide; Stipe: 3–10 cm long and 2–6 cm wide

Morels look like giant shrivelled sultanas once they're cooked. But do not be deceived: these magnificent mushrooms pack a flavourful punch, and have a colourful history too. They're meaty, chewy, and full of umami depth. My grandmother, who was raised in the foothills of North Carolina, USA, had a friend who hunted morels. They always said they'd take her along, but never did; morel spots are more precious than gold.

While these strange mushrooms have been foraged in the United States for generations, they aren't as well known in Aotearoa. They are somewhat rare, but each year it seems that more and more people find them. I'm not sure if there are more morels, more hunters, or both. Either way, these delicious morsels are out there.

In Aotearoa, morels exist in a somewhat grey area. We know we have *Morchella importuna*, the black landscaping morel, but more taxonomic work needs to be done to figure out if other edible species are present. We also have native morels here that haven't yet been described.[9]

Another thing that makes morels so intriguing is their elusive nature. They're hard to grow commercially and aren't regularly imported. You've got to find them, or buy them from a speciality shop, or grow your own — and if you're keen to try this 'dark art', you can buy morel spawn online within Aotearoa. It's a tricky endeavour, with no guaranteed success. But if you've got a mulch bed and some space, it could be worth a shot.

How to find and identify

Environment: Tends to grow in woodchips on disturbed ground, and recently burned areas. As the common name suggests, they like habitats that humans have created.

Cap: Not a classically capped mushroom — the honeycomb-like texture found on the upper part of this fungus is unique. Its ridges are dark and its pits are lighter in colour. The cap is conical, and completely hollow inside when sliced open.

Spore surface: Morel spores are held in the pits of the honeycomb texture.

Spores: Cream-coloured.

Stipe: Light in colour, and widens at the top to join the cap. Both the cap and the stipe should be completely hollow when sliced open.

Dangerous look-alikes: Be aware of the genus *Gyromitra*, the false morels. These have a similar shape but their caps do not have a consistent honeycomb texture — they're more lumpy and look more like a brain or shrivelled sack. They are also not completely hollow inside. False morels can be deadly poisonous.

How to use

Morels don't need a lot of help. They're best cooked with butter and a bit of salt and pepper. Stir them through pasta with peas and a bit of cream for a super-memorable meal.

Black truffle

Scientific name
Tuber melanosporum

Other common names
Perigord truffle, French black truffle

Typical environment
Truffle farms

Difficulty
Hard

Distribution
Mostly on truffle farms, but some wild truffles have been reported in Canterbury

Peak season
Autumn into winter

Size
Can grow up to 10 cm across

Truffles are the crown jewel of edible fungi. They have been enchanting palates, beguiling growers and inciting everything from excitement to violence for centuries. Two things make truffles unique — they grow underground, and they smell. These two attributes are inextricably linked. Truffles and mushrooms have the same goal: to reproduce. Yet truffles never emerge from the subterranean to drop spores. Instead, they remain hidden away and rely on their intense scent to get the job done.

The truffle funk is an incredibly clever adaptation. It's powerful and attention-grabbing. On the forest floor, energy-conscious scavengers will stop in their tracks and dig to find the source. The scent also enraptures humans. More than a few have dedicated their lives to growing these mycelial morsels. Gareth Renowden, originally from Britain, helped establish the truffle industry in Aotearoa. Of course, growing truffles isn't easy, Gareth taught me. They grow in association with certain plants, so to 'grow' a mycorrhizal fungi like this you've actually got to plant trees that have been inoculated with truffle spores. Then, you've got to wait. And train a dog in the meantime. Gareth's dog Rosie, an adorable beagle, uses her nose to help Gareth figure out if truffles are there, and if they're ripe.

There are a few reports of wild truffles here in Aotearoa. These could be 'weed' truffles that have escaped from truffle plantations and spread with the help of animals. Or they could be the work of humans — more than a few have tried to set up their own small patches but abandoned them over time.

How to find and identify

Environment: Black truffles in Aotearoa are almost exclusively found on truffières (truffle farms), where they're grown in association with oaks and other deciduous trees; many farms offer tours and guided hunts in autumn. As for the wild truffles that have been reported, more taxonomic work is needed to determine what species they are.

Fruiting body: Rounded and blob-shaped. The outer texture is black, rough and grooved, almost like the bark of a tree. The interior has a marbled look to it.

Scent: Unlike anything else. A truffle simply smells like a truffle. But it has been described as musky, garlicky and deep.

Look-alikes: Rhizopogon species, the 'false truffles', could be confused with true truffles. *Rhizopogon* species are usually tan or yellow and resemble a potato on the outside. On the inside, they start white but turn olive as they mature. They are not aromatic or edible.

How to use

If you've got a truffle, treasure it; and remember that its flavour goes best with a fat. Get a wheel of brie, cut it in half horizontally, put a few slices of truffle in the middle, return it to the fridge and let it sit overnight. Then make a truffle brie toastie. You can also put a bit of truffle in with your eggs or butter. They will take on the truffle flavour just by being in proximity with it — that's how powerful this evolutionary aroma is.

5

Fungi of Aotearoa

Birds that don't fly, loads of sheep, some humans, and absolutely no snakes — everything in Aotearoa is just a little bit different. And that's true of our fungi, too.

While you could spend a lifetime trying to spot a kiwi, our endemic fungi are a touch easier to locate. For starters, they can't move. Well . . . one species can; but I'll come back to that. The fungi of Aotearoa are wonderful, wacky and downright weird. On a day-walk in Arthur's Pass, I once spotted fungi in every single colour of the rainbow. Red, orange, yellow, green, blue, indigo and violet. In just six hours.

These colours, of course, aren't just for show — there's a story behind every striking shade. Take our native truffle-like fungi, for example. Their lilac, lapis and lime fruiting bodies may have evolved specifically for moa. These vibrant colours sing out from the forest floor. They say, 'Hey! Giant bird! Look down here. Come eat me. Spread my spores!' And where does this hypothesis come from? From moa poo fossils, which, in a more polite sense, are called coprolites.[1]

Like our flora and fauna, the fantastic fungi of Aotearoa are undeniably funky. New Zealand's geographical isolation has allowed them to write their own evolutionary tales. Like teens hell-bent on being different, our fungi have quite literally gone their own way. The end result? Some of the strangest, most out-of-this-world organisms you could ever hope to meet.

Previous page Pink waxgill mushrooms.

Another thing to keep in mind as you fall down this rabbit hole: fungi are tough. They have been launched into space and survived the journey. They can bring entire ecosystems back to life after volcanic eruptions. They even dabble in mind control. Fungi have endured, and will endure. Despite this, many of them are threatened. Over 50 species of Australasian fungi are critically endangered, and more work is needed to describe and preserve the biodiversity we have.[2]

As I've discussed throughout this book, our fungi are stunning, specific and key to our survival. Looking for them, identifying them, caring about them — this is more than just a hobby. When you go fungi-spotting, you're observing and appreciating a global public good. So let's go meet the Black Ferns and All Blacks of the underground. The most striking fungi this great nation has to offer.

How to use this section

Aotearoa is home to around 22,000 species of fungi — that's our mycologists' best guess. It's impossible to know exactly how many species we have. But we do know one thing for sure — there's a whole heap of cool fungi here. This section is a guide to our most common and most intriguing macro-fungi. It will help you get familiar with, and is organised by, the key 'groups' of fungi that are found here.

In some groups, all the species featured are part of the same genus. The webcaps, for example, are all within the *Cortinarius* genus. Other groups, though, like the jelly fungi, are organised by a key feature they share, like texture or shape. From classically capped *Amanita* mushrooms, to cup fungi, lichens and jiggly jelly species, there's a lot to explore. Within each group I've included the most notable species as well as quick references to others you might spot within that group.

With an understanding of these groups, you'll have a broad view of what's here and will be able to place your finds into their larger group/genus.

For each species, I've given the common, scientific and Māori names where possible. There are many dialects of te reo Māori, though, so I do want to note that not all names may be captured here. In terms of scientific names, I've named to the species level where possible. In some instances, we know what we have in Aotearoa is distinct to similar species overseas, but taxonomic work to show this and give a unique name to what *we* have has not yet been completed. In these cases I've written the scientific name as '*Genus species* group'. I've given a general indication of when and where each species may be found in Te Ika-a-Māui North Island and Te Waipounamu South Island (shortened to NI and SI). This info comes from iNaturalist data. Size information has come from a variety of existing publications, all listed at the back of this book. And further info has come from kōrero, other guides and my own observations.

As you go along, remember this — probably fewer than half of our fungi have been described so far. There is opportunity to describe your very own species. The process takes a while, but the humble fungi-spotter, your average citizen scientist, can play a serious role in documenting, describing, and learning more about this world. Uploading your finds to the iNaturalist website is a great place to start.

A fallen *Mycena ura*.

A pink *Entoloma* species.

Pinkgills

The pinkgill group, genus *Entoloma*, contains some real beauties, and our national fungus too. We've got a grey one with a fruity smell. A brilliant blue one, which features on our $50 note. A lovely peachy orange one, a pale lime-green one, and a brown one that looks like a bullseye from above. Over 75 *Entoloma* species have been described here, and 1000 species globally. Their colours, peaked caps and slightly striated stipes make them unique on the forest floor.[3]

Entoloma mushrooms are sometimes referred to as pinkgills as they have pink-coloured spores. While many pinkgills grow from the leaf litter and are mycorrhizal, some are saprobic and grow from rotten tree stumps.

Werewere-kōkako

Scientific name
Entoloma hochstetteri

Other common names
Blue pinkgill, sky-blue mushroom

Distribution
Found across the NI except in Gisborne and Hawke's Bay. In the SI it grows in Nelson, Tasman, Marlborough, West Coast, Arthur's Pass and Southland

Peak season
February to June

Size
Cap: 2–5 cm across; Stipe: up to 5 cm long

Werewere-kōkako was voted our national fungus in 2018, and features on our $50 note. They are elusive in some areas, more common in others, but always absolutely stunning to spot. I spent almost an hour gazing at the first one I found. Look for them in dense native bush. They like to be shielded from the scorching sun and cradled by pillowy beds of moss. If these fungi were human, they would only stay at luxury eco-lodges while travelling.

Check out the $50 note and you'll also find the North Island kōkako, a native bird that is quite striking. Both organisms can be found in Te Urewera, the dense, misty rainforest that is home to Tūhoe iwi. The whakataukī story of Tūhoe is that the kōkako bird was in a bit of a hurry. As he darted through the bush, he brushed his cheek against this mushroom, took a bit of the pigment with him, and that's how he got his blue wattles.[4]

Jade pinkgill

Scientific name
Entoloma glaucoroseum

Distribution
Has been spotted in the Bay of Plenty in the NI, and in the West Coast and Canterbury regions of the SI

Peak season
April to October

Size
Cap: 1.5–2.5 cm across; stands up to 4 cm tall

This species is a green goddess. She has creamy white gills, which become pink over time, the colour of pounamu on her cap and a stipe that fades from orangey-brown at the base to pale green towards the top.

These caps tend to start out with a convex shape, but as they open they flatten and fan out, and sometimes get a bit depressed or dimpled in the centre. Look for this species in the cooler months in podocarp forest among trees like kauri, miro, tōtara and kahikatea.

A new blue

In nature, blue organisms are rare. The chemical compounds that produce blue hues are extremely sensitive to oxygen. They become unstable in its presence and quickly turn brown. As a result, most blue 'food' is coloured with artificial dye. Werewere-kōkako somehow retains its unreal shade over time, though. Food chemist Silas Villas-Boas is using sophisticated genetic technology to try to isolate the mushroom's blue genes, and replicate them to produce the world's first stable, yet nearly neon, natural blue food dye. If he is successful, Aotearoa could become home to the holy grail of food chemistry — an all-natural blue food dye that doesn't pale (literally) in comparison to artificial options.

Werewere-kōkako is a taonga species, so working with it in a potentially commercial context, Silas taught me, requires genuine partnership with iwi and hapū. In keeping with Te Tiriti o Waitangi and the Wai 262 Claim, continued collaboration, consent and acknowledgements will help ensure Māori retain authority over this taonga and its genetics if this work continues and is successful.

Entoloma haastii

This large *Entoloma* species has a deep, dark, moody-blue cap, creamy whitish gills that turn more pink over time, and a stipe that features swirls of blue and white towards the bottom. Look for it under beech trees and in broadleaf podocarp forests, often nestled among leaf litter and moss. Only found in Aotearoa.

Common names
None yet

Distribution
Auckland, Bay of Plenty, Manawatū and Wellington regions in the NI; West Coast, Arthur's Pass and Southland in the SI

Peak season
February to July

Size
Cap: 1.5–5.5 cm across; Stipe: 4–10 cm long

Entoloma perzonatum

These little bull's-eyes have concentric rings of buff and brown, and a distinctive felted wool texture. Young ones, with their rounded caps, almost look like sheep droppings from above. It is the only fungus I know of that features tidy stripes. So unique. Look for it growing among and tucked beneath grasses in more open environments (these guys can survive in fairly exposed areas), and nestled among mosses in more sheltered environments.

Common names
None yet

Distribution
Found in the NI south of Taupō, and across the SI apart from Marlborough

Peak season
October to May

Size
Cap: 2.5 cm to 4 cm across

Entoloma canoconicum

Common names
None yet

Distribution
Found across the NI apart from in Gisborne, and the West Coast, Southland and South Otago in the SI

Peak season
February to July

Size
Cap: up to 4 cm across; stands up to 5 cm tall

This species is endemic to Aotearoa. You can only find this wee beauty here. It has a light-grey to light-blue colour, and a distinctive cone-shaped cap. The cap sometimes takes on a glistening quality after rain, or if the area is particularly damp. The gills are a similar colour to the cap. Look for this species in the cooler months under mānuka trees and in broadleaf podocarp forests.

Entoloma latericolor

Common names
None yet

Distribution
Found in the NI north of Taupō

Peak season
February to July

Size
Cap: up to 4 cm across; stands up to 8 cm tall

Like many other *Entoloma* species, this one has a cone-shaped cap that sometimes gathers into a distinctive point at the top. The cap is an orangey-brown colour and can be slightly fuzzy. Like a few other *Entoloma* species, the stipe sometimes has faint stripes, which almost make it look twisted or braided together. The stipe is lighter in colour than the cap, and the gills are a rusty-orange colour but become more red with age. Look for this species under kauri trees and in broadleaf podocarp forests.

Bonnets and helmets

The *Mycena* genus features teeny tiny, hat-shaped species in a huge range of colours, from a shimmery ice-blue to blood red, stark white, lime-green, pastel pink and magenta. Most *Mycena* caps are just 5 to 15 mm across, so we're talking really small here. These mini-mushrooms are difficult to snap a photo of without a professional camera set-up. They are usually just for enjoying in real time.

The distinctive cap shape of *Mycena* mushrooms makes them look like colourful headwear for fairies. Some *Mycena* species are bioluminescent — so make that colourful, glow-in-the-dark headwear. In 2021, fungi-hunters on the annual New Zealand fungal foray were the first to observe that our native pink helmet, *Mycena roseoflava*, also glows in the dark.

In the past, bioluminescent fungi were known as foxfire. Over 2000 years ago, Aristotle made the first mention of foxfire, and Pliny the Elder also mentioned 'glowing olive wood' in his records. Foxfire was used to help light barometer needles and other directional tools on early submarines.

Mycena mamaku.

Crimson helmet

Scientific name
Mycena ura

Distribution
Found up and down Aotearoa, with a sighting in nearly every region

Peak season
March to July

Size
Cap: 3–7 mm across; stands up to 60 mm tall

These blood-red beauties are endemic to Aotearoa. I first spotted this species as I was finishing the Tongariro Alpine Crossing. The last section of that tramp crosses a lahar flood plain. A lahar is what results when heavy rainfall hits the side of a volcano and causes mud and debris to tumble down into the valley below with devastating power. You're meant to move quickly through this area, so I paused for just 60 seconds to admire these tiny caps. I was amazed that something so small could be so eye-catchingly bright. This species even bleeds a red sap when broken. Look for it on downed wood in forests all over Aotearoa.

Pixie's parasol

Scientific name
Mycena interrupta

Distribution
Grows in every region apart from Northland and Central Otago

Peak season
April to June

Size
Cap: 8–12 mm across; stands just 30 mm tall

There's something extra-special about blue fungi. And second only to werewere-kōkako, this one was a Holy Grail to me when I first started looking for fungi.

I cannot overstate how tiny this species is. It took me three years of training, three years of bush-walking with fungi in mind, to spot this one, and I finally spied it on a tramp on the West Coast. This species grows on rotting logs and has a brilliant blue colour in the centre of its cap, which fades towards the edges. It's almost sparkly and has a distinctive pad structure at the base of its stipe. This fungus follows a Gondwanan distribution pattern, meaning that it's found here as well as in Australia, New Caledonia and Chile — all bits of land that touched one another before the Gondwanaland supercontinent broke up 180 million years ago.

Mycena roseoflava

Common names
None yet

Distribution
Scattered throughout Auckland, Hamilton, Tauranga, New Plymouth and Napier. More concentrated in the Wellington and Manawatū regions. Also found in Nelson, Tasman, Marlborough, West Coast, Canterbury, Southland and around Dunedin

Peak season
March to July

Size
Cap: up to 10 mm across; stands 5–10 mm tall

A tiny, bubble-gum-pink mushroom that's native to Aotearoa. The stipe is a pale, whitish yellow. This species was first described by Greta Stevenson, a New Zealand botanist and mycologist, in 1964. In 2021 it was found to be bioluminescent. Look for it by day on rotting twigs and logs. Return at night to witness its glow: turn off headlamps or torches and give your eyes plenty of time to adjust in order to see it.

Yellow-legged helmet

Scientific name
Mycena subviscosa

Distribution
Found in most regions of the NI apart from Gisborne. Sightings in the SI are concentrated in the northern regions and the West Coast, with a few sightings around Dunedin and Southland

Peak season
April to July

Size
Cap: 5–10 mm across; stands up to 60 mm tall

Very small, slightly sticky on top, and with a yellow stipe, this species can be found in social groups on downed branches and logs. It's native to Aotearoa and some say it has a cucumber-like smell. While I've yet to find this one, or test its scent, the colour palette this species sports does lend itself to a cucumber feel. Look for this species in the cooler months.

Mycena flavovirens

Common names
None yet

Distribution
Found in a narrow band in the NI from Auckland to Te Urewera

Peak season
March to August

Size
Cap: up to 10 mm across; stands up to 30 mm tall

This species looks like the green elixir of the fungi world. It's been found, but only a few times, on rotting ponga logs. The cap is dark green, and the entire body of the mushroom fades out from there, going yellow towards the base. This globally threatened species has only been sighted about 14 times in the past 50 years. Keep your eyes peeled for this one in high-rainfall forests among rotting tree ferns.

Mycena mariae

Common names
None yet

Distribution
Found across the NI and SI apart from in Central Otago and South Canterbury

Peak season
April to July

Size
Cap: up to 20 mm across; stands up to 55 mm tall

This species is endemic to Aotearoa. Its small caps are a light pink colour with a darker red shade at the centre. The stipe is a darker pink and bleeds a red 'sap' when damaged or broken. Look for this species on dead wood or decaying twigs, and within leaf piles.

Gliophorus versicolor.

Waxcaps and waxgills

This group shares a common texture. They're all brittle, break easily, and feel waxy when crushed between your fingers. I can't imagine a scenario where you'd want to crush one of these beauties, though.

The two main genera in this group are *Hygrocybe* and *Gliophorus*, or waxcaps and waxgills, respectively. They're all semi-small soil-dwellers that pop up in the colder months. Some species within this group are so bright that they look like Christmas lights amid the chilly winter bush. They're cheerful even when the weather doesn't want to be.

Waxcaps (*Hygrocybe*)

Scientific name
Hygrocybe species

Distribution
Found across Aotearoa

Peak season
March to July

This genus name offers helpful clues for identification. 'Hygro' means wet, and 'cybe' means cap or head. These mushroom caps are usually cool, a bit damp and waxy to the touch, even though they're fiery red, orange, yellow or crimson in colour. They're like little bonfire beacons on the forest floor; perfect to curl up next to if you're a fairy.

Over 20 waxcaps have been described in Aotearoa. I often see tiny ones while out tramping. They seem to like the strip between the trail and the bush where there's not too much plant life and the soil has been disturbed. Look for native species in the bush and introduced ones in urban areas. Some waxcap species live in association with moss, so look there too.

Waxcaps are some of my favourite fungi to photograph. They offer a huge range of colours, and some species have frilly, semi-upturned edges. This makes it easy to get a photo of both the profile and underside of these magnificent mushrooms in one shot. I've also noticed that if I see one waxcap species, there's usually a few others close by.

Hygrocybe procera.

Goblet waxcap

Scientific name
Hygrocybe cantharellus

Distribution
Northland, Bay of Plenty, Wellington in the NI; Nelson, Tasman, West Coast, Canterbury, Southland in the SI

Peak season
December to June

Size
Cap: 6–20 mm across; Stipe: 30–70 mm long

Native to Aotearoa, and features a bright orange colour on the cap with paler gills underneath. The gills run up the stipe and out to the edges of the cap; from a side view this fungus looks like a fancy drinking vessel. It usually pops up in mossy, boggy places.

Vermillion waxcap

Scientific name
Hygrocybe 'miniata'

Distribution
Found across the NI apart from in Gisborne and Taranaki, and in Nelson, Tasman, West Coast and Southland in the SI

Peak season
March to August

Size
Cap: 20–35 mm across

This native species is small, but makes up for this with colour. Its brilliant red cap and stipe are hard to miss, as are the contrasting white-yellow gills underneath. Some caps fade to yellow at the edges. Look for this species hiding beneath tree ferns or growing from mossy beds in most forest types.

Rare waxcap

Scientific name
Hygrocybe rubrocarnosa

Distribution
Auckland, Waikato, Bay of Plenty, Wellington in the NI; West Coast, Southland and around Dunedin in the SI

Peak season
March to August

Size
Cap: up to 40 mm across

Not rare to spot, but rare-looking when it's cut open — like a steak. The cap is red with a slight bronze sheen to it. Sometimes slugs and snails nibble on the top, revealing the rare red colour within. Look for this one in native bush, among damp leaf litter. It's endemic to Aotearoa.

Hygrocybe julietae

Common names
None yet

Distribution
Auckland, Bay of Plenty, Wellington in the NI; Nelson, Tasman, West Coast, Southland and around Dunedin in the SI

Peak season
April to August

Size
Cap: up to 20 mm across; stands up to 50 mm tall

A somewhat common but still super-striking waxcap — a burst of yellow that can be found in most forest types. The edge of the cap is a bit frilly and festive. Look for it among leaf litter in lowland broad-leafed podocarp forests.

Blackening orange waxcap

Scientific name
Hygrocybe astatogala group

Distribution
Found throughout the NI, and in Nelson, Tasman, Marlborough, West Coast, Arthur's Pass, Southland and around Dunedin

Peak season
January to August

Size
Cap: up to 35 mm across; stands up to 100 mm tall

This species has a bright, conical yellow-red cap, which has unique black 'hair'; these tufts of texture are called fibrils. The stipe is usually a deep shade of red. Look for it among leaf litter.

Witch's hat

Scientific name
Hygrocybe conica group

Other common name
Conical slimy cap

Distribution
Found across Aotearoa

Peak season
February to June

Size
Cap: up to 40 mm across; stands up to 100 mm tall

This species is fairly common. When it first pops up, it has a cone shape and a bright yellow-red colour that is often sticky to the touch. When bruised and with time, this species turns black, hence its common name. This species is more witchy than fairy! Look for it on your lawn or growing on berms among the grass.

Hygrocybe firma, a blood red waxcap that grows among leaf litter.

Waxgills (Gliophorus)

Scientific name
Gliophorus species

Distribution
Found throughout Aotearoa

Peak season
April to August

The common name for waxgills ought to be changed to 'nightlights for fairies'. Waxgills are often coated in a slimy, gelatinous texture. It sounds gross but it amplifies their colours and gives them a shimmery, shiny quality. If you squint a bit, they look like they're made of frosted glass.

Waxgills often grow in large groups. Spot one and you might spot ten, or even a hundred, more. The first time I saw waxgills, I saw not one species but about four in baby pink, pale yellow, pea green and light lilac. Together they looked like a whimsical bag of pastel pick 'n' mix.

In my experience, waxgills are a special find only happened upon in dense native bush sections that aren't easily accessed from my city abode. This makes them extra exciting to spot. Because they're usually tucked away in the bush, where the sun only just gets through, these can also be hard to get a good photo of. In that way, they're almost like fungal snowflakes. Hard to capture, absolutely spectacular to look at and difficult to describe.

Gliophorus versicolor.

Verdigris waxgill

Scientific name
Gliophorus viridis

Distribution
Found throughout the NI, and in the Tasman, West Coast and Southland regions

Peak season
April to August

Size
Cap: 15–25 mm across; stands up to 50 mm tall

This species is gorgeous, and oh so very green. It's native to Aotearoa and springs up from similarly green substrates like cushiony moss, or soft leaf litter on the forest floor.

Gliophorus chromolimoneus

Common names
None yet

Distribution
Found throughout most of the NI, as well as Nelson, Tasman, Marlborough, West Coast, Canterbury, Southland and Dunedin

Peak season
March to August

Size
Cap: up to 30 mm across; stands up to 50 mm tall

A highlighter-yellow, you-can't-miss-me kinda colour — that's this species' calling card. Look for it in the cooler months in the leaf litter beneath silver beech trees.

Slimy green waxgill

Scientific name
Gliophorus graminicolor

Distribution
Found throughout the NI, and in Tasman, Marlborough, West Coast, Southland and around Dunedin in the SI

Peak season
April to August

Size
Cap: up to 20 mm across; stands up to 45 mm tall

Another incredibly green species. Slightly different to the verdigris waxgill, though — this one has a slimy thread right to the edge of its gills. Plus, it has a frosted-glass-like look to it.

Rose waxgill

Scientific name
Gliophorus versicolor or *Gliophorus ostrinus*

Distribution
Found around Auckland, Hamilton, Tauranga, Rotorua and Wellington in the NI, and Nelson, West Coast and Southland in the SI

Peak season
April to December

Size
Cap: up to 20 mm across; stands up to 50 mm tall

A precious pink colour, these grow in and among leaf litter in many forest types. Common in some areas, but pretty special.

Gliophorus lilacipes

Common names
None yet

Distribution
Northland, Auckland, Waikato, Manawatū and Wellington in the NI. Nelson, Marlborough, West Coast and Southland in the SI

Peak season
April to August

Size
Cap: up to 20 mm across; stands up to 60 mm tall

Similar to the rose waxgill, but more lilac in colour. This species is a bit of a chameleon, and can also be quite blue. Look for it in podocarp forests.

Gliophorus pallidus

Common name
None yet

Distribution
Auckland, Bay of Plenty, West Coast of the SI and Southland

Peak season
April to July

Size
Cap: usually 15–20 mm across

This mushroom's colour reminds me of golden kiwifruit. The colour tends to fade to white with age and it has a glutinous texture. Look for this species growing in beech and podocarp forests from dead wood.

Gliophorus viridis, of a different shade.

Laccaria

Native species within the genus *Laccaria* can be found growing in association with beech and mānuka trees. Introduced *Laccaria* species grow in association with pine, oak and birch trees, among others.[5] They come in some pretty cool colours, and sometimes look a bit like waxcaps (*Gliophorus* species) but don't have the same 'wet' texture.

There's some evidence that *Laccaria* species may be ecosystem pioneers — the brave souls who emerge first after a tract of land has been stripped for forestry, for example. They're often found in tree nurseries and on disturbed sites.[6] I often spot them in the Barbadoes Street Cemetery in Ōtautahi Christchurch.

In addition to their violet, pink and ochre hues, the gills of *Laccaria* species have a fun vibe about them. They're wavy and slightly disorganised. Very free and happenstance, like the happy-go-lucky friend who's never too worried about colouring in the lines.

Laccaria masoniae.

Laccaria masoniae

Common names
None yet

Distribution
Manawatū and Wellington regions in the NI, and Nelson, Tasman, West Coast, Arthur's Pass and Southland in the SI

Peak season
April to August

Size
Cap: up to 25 mm across; stands up to 100 mm tall

This *Laccaria* species is unique not only for its lovely purple hue, but also for how it ages. Over time, the purple fades to tan. Since this species seems to emerge in clusters, you might find a heap of bright lively purple ones interspersed among slightly older, wiser brown ones.

This species may also be an 'ammonia fungus', as it's frequently found growing among bones and decaying animals and in soils that are rich in nutrients. Its hues remind us of the stages of life, and so too does where it chooses to grow.

The deceiver

Scientific name
Laccaria laccata

Distribution
Northland, Auckland, Bay of Plenty, Gisborne, Hawke's Bay and Wellington regions in the NI. Nelson, Tasman, Canterbury and Otago in the SI

Peak season
March to July

Size
Cap: up to 30 mm across; stands up to 100 mm tall

The common name of this introduced species relates to its variability: its colour ranges from washed-out to reddish brown and even orange. It is sometimes considered a 'weed mushroom' because it's so common on lawns and verges. Look for it in grassy areas under exotic trees like oaks and birch.

Saffron milkcap.

Milkcaps

Alternative milks are everywhere — oat milk, macadamia milk, even potato milk. But what about . . . mushroom milk?

Cut or bruise a mushroom in the milkcap genus *Lactarius*, and you'll find it exudes a strange white substance called latex. It seriously does look like milk. Lots of plants do this, actually. If you've ever grown lettuce at home and cut the head off at the base, you might've noticed that it puts out a white substance too. About 10% of plants produce latex.

And why do plants and fungi produce this strange milk? Its bitterness might deter hungry insects, and its stickiness might trap the mouth parts of those insects, halting their onslaught. It might also function as a sort of protective seal after injury. Whatever the reason, it's pretty cool to spot these milky mushrooms in the bush.

Look for native milkcaps beneath beech and mānuka trees, and look for introduced milkcaps growing in association with silver birch and pines. You'll know you have a milkcap when you notice the latex oozing out. The cap's slightly dimpled centre is also a dead giveaway.

Coconut milkcap

Scientific name
Lactarius glyciosmus

Distribution
Found around Palmerston North and Wellington in the NI and around Christchurch, Dunedin, Queenstown, Wānaka and Dunedin in the SI

Peak season
March to May

Size
Cap: up to 8 cm across; Stipe: 2.5–6.5 cm long

While I haven't yet had the pleasure of finding this one, I eagerly await the day. It's a milkcap that apparently smells of coconuts. Which begs the question . . . is this fungus edible?

Most mycologists consider this species inedible, but there are reports of it being sold in rural fungi markets in China. I don't think I'll be trying it anytime soon, but it is a good reminder that some fungi are considered edible — even delicious — in one part of the world, and inedible in others. The caps of this species are greyish with mauve and brown undertones, and the latex it exudes is white.

Look/sniff around for this one under silver birch trees.

Ugly milkcap

Scientific name
Lactarius turpis

Distribution
Waikato, Manawatū and Wellington regions in the NI, and in Nelson, Marlborough, Canterbury, Otago and Southland in the SI

Peak season
February to June

Size
Cap: 8–20 cm across; Stipe: about 7 cm long and 3 cm wide

This brownish-green milkcap can grow quite huge — up to 200 mm across! It's usually found under silver birch and other introduced trees, since it is also an introduced species, and it has a girthy stipe. It tends to collect debris from the forest floor on its slightly sticky cap. While this common name isn't very kind, this species is still cool to spot. Younger specimens sometimes have a velvety texture or shaggy rim.

Downy milkcap

Scientific name
Lactarius pubescens

Distribution
Auckland, Waikato, Manawatū and Wellington regions in the NI and Nelson, Tasman, Canterbury, Otago and Southland in the SI

Peak season
February to June

Size
Cap: 2.5–10 cm across; Stipe: 2–6.5 cm long

This species is one I see all the time in my urban fungi-spotting adventures; it often grows alongside silver birch trees. Its light-pink, downy cap is nice to pat, and the way it rolls over the edge, wrapping the gills in a woolly embrace, is a unique feature. The edibility of this species is both controversial and ambiguous. It is consumed in Russia after a lengthy and delicate marinating process to draw out toxins, but it is generally not recommended for consumption. A fungus I love to see, but not one I'll be eating anytime soon!

Saffron milkcap

Scientific name
Lactarius deliciosus

Distribution
Grown on fungi farms; may escape into surrounding areas

Peak season
April to June

Size
Cap: 4–14 cm across; Stipe: 3–8 cm long

While there a few sightings of this one in the bush, it's mainly just found on fungi farms. There are a few growers of this species up and down the country who cultivate it under introduced pines. Saffron milkcap cultivation has also been trialled in commercial pine plantations. This species has a carroty-coloured top, and its latex is also orange, but, oddly, it stains green when broken or bruised. To spot this one, or even try it yourself, hop online. You can order this tasty fungus delivered direct to your door.

Hare's foot inkcaps.

Inkcaps

The inkcaps are best known for what they become: inky black pools of goo. They can actually be used to make ink. Collect one, give it a few hours to fully liquefy, mix it with a bit of water and grab a brush. Voilà! You're painting with fungi.

The inkcaps sit across several genera (genuses) and vary in size, shape and texture. The shaggy inkcap, *Coprinus comatus*, which you can read about on page 120, is a choice edible species. They're fairly easy to identify and cook with, if you can find them before they go inky. Inkcaps are also fun to know about because, at times, they are seemingly everywhere; they are some of the easiest fungi to spot and identify. They love areas where people spend heaps of time — city blocks, parks, berms, your garden, and so on.

Hare's foot

Scientific name
Coprinopsis lagopus group

Other common name
Revolute inkcap

Distribution
Found across Aotearoa

Peak season
March to June

Size
Cap: 3–4 cm across; stands 5–18.5 cm tall

These inkcaps often pop up in clusters and have elegant, long, almost luminescent stipes. When they first emerge their caps are conical, grey and have white tufts on them. Then they quickly expand and turn upwards. At this stage they look like a tiny, cheap umbrella turned inside out by a Wellington gust. These last only a few hours.

Mica cap

Scientific name
Coprinellus micaceus group

Other common name
Glistening inkcap

Distribution
Auckland, Waikato, Manawatū, Wellington Nelson, Tasman, Canterbury and Otago

Peak season
April to July

Size
Cap: 1–2.5 cm across; Stipe: 3–10 cm long

These inkcaps are fairly common, and the little white granules on their caps make them highly recognisable. You can brush these away with your finger, like fairy dust. The caps usually have a light yellow-brown colour to them. The stipes are frosty white. Very common in my garden, the park and other urban areas.

Fairy inkcap

Scientific name
Coprinellus disseminatus group

Other common name
Trooping crumble cap

Distribution
Found throughout the NI and in Nelson, Tasman, West Coast, Canterbury and Dunedin

Peak season
March to June

Size
Cap: 0.5–1.5 cm across; stands 1–1.5 cm tall

These are also known as the 'social inkcaps' as they tend to cluster together in giant troops. I've seen hundreds of them taking over a single rotting log, and there's something stunning about their careful yet semi-disorganised repetition. This species doesn't dissolve into ink, but does have 143 different mating types, or rather, sexes.[7]

Common inkcap

Scientific name
Coprinopsis atramentaria group

Distribution
Found throughout Aotearoa

Peak season
February to October

Size
Cap: 3–10 cm across; Stipe: 5–17 cm long

Keep an eye out for common inkcaps on lawns and along paths. They have shiny, brownish bell-shaped caps that flatten out as they age and darken. The stipe is white. These are sometimes also called 'tippler's bane', as they are technically edible but will make you sick if alcohol is consumed within two days either side of eating. Some reports say that even using an alcohol-based aftershave after eating these can make you vomit. I don't mess with this one.

Splitgills, toughshanks and wood-knights

The splitgills, toughshanks and woodknights sound like feuding rugby clubs. But really, these are three stunning fungi groups. They all have beautiful gill structures yet hardy forms, and can be found on downed branches and twigs.

Let's start with the splitgills. They all fall into one genus, *Schizophyllum*. While there are only six splitgill species globally, and only one is found in Aotearoa, this genus may be the world's most widespread fungus. So wherever you happen to be fungi-spotting, you can be semi-sure these are around.

The toughshanks have deeply grooved caps with scalloped edges. Despite their tiny size, their stipes are wiry and hard to break. Their gills are widely spaced and wandering; sometimes one seems to flow into another.

Last but not least, spiny woodknights are covered in unmistakable orange spikes.

A spiny woodknight.

Splitgills

Scientific name
Schizophyllum commune

Distribution
Found throughout Aotearoa

Peak season
Always about

Size
Each fruiting body is about 20–30 mm wide

Splitgill fungus is tough yet delicate, like your grandmother's 100-year-old lace doily collection. They are leathery yet slightly fuzzy. Their borders flow out from the central point of connection to whatever piece of dead wood they are growing on, and are wavy and non-uniform. The gills underneath split away from one another in intricate patterns.

When you do see one, and you eventually will, go see it again after a big rain or a dry spell. These fungi are tough-as. They can curl up and survive long periods without rain, only to unfurl to their full glory again after a storm passes through. This species is, remarkably, found on every continent except Antarctica; it can survive just about anywhere. In one study, a similar species sprang back to life with the addition of water after being vacuum-sealed in a bag for 35 years![8]

This species also has over 23,000 different mating types, which can be thought of like sexes.[9] And its name holds a few clues about its personality, too. *Schizophyllum* is derived from the Greek word 'schiza', which means split, like its radially expanding, diverging gill patterns. And *commune*/common indicates that they're everywhere, and often growing alongside one another.[10]

Toughshanks

Scientific name
Crinipellis species

Other common names
Pinwheel and parachute mushrooms

Distribution
Found throughout Aotearoa

Peak season
Always about

Size
Cap: 10–15 mm across; stands up to 150 mm tall

There are at least a handful of toughshank species in Aotearoa, but they are not yet well described. They usually have brown to white caps and a dark stipe. The stipes are thin and sometimes long and elegant yet still quite strong, almost like little bits of wire. From above, these caps look like pinwheels with their lobed, scalloped edges.

While these species are often found growing on fallen twigs and branches, I have seen extremely tiny toughshanks growing from a single leaf in a cluster of four or five. Each cap must have been 1 mm or less across. Tiny, delicate, magic; yet also sturdy. Keep an eye out for toughshanks in damp sections of the bush where there is plenty of dead organic matter for them to eat.

Spiny woodknight

Scientific name
Cyptotrama asprata

Other common name
Golden scruffy collybia

Distribution
Found in the NI north of Taranaki

Peak season
November to July

Size
Cap: up to 30 mm across; stands up to 40 mm high

This species can be found on dead wood. It has a scruffy, softly spiked cap. The cap and the stipe are bright orange-yellow and the gills are white. This species' spiny armour is iconic, and unlike any other species found in Aotearoa. Look for it on decaying wood in broadleaf podocarp forests.

Fly agaric.

Amanitas

The *Amanita* genus includes some heavy hitters. Within it you'll find the classic red and white toadstool as well as some of the world's most deadly poisonous species. At least 10 of the 20 or so species found here are native to Aotearoa, but there are over 600 globally.[11]

Amanita species are most easily identified by the warts/scales/patches they sport on their caps. Many of these species get quite large, have a pronounced skirt on their stipe and a swollen base. Some grow from a volva. *Amanita* species all have white spores.

Most *Amanita* species are mycorrhizal. They spring forth from subterranean mycelial networks that commingle with plant roots. Introduced *Amanita* species tend to associate with introduced exotic trees like pine and birch; native *Amanita* can often be found underneath beech and mānuka trees.

Fly agaric

Scientific name
Amanita muscaria

Distribution
Found throughout Aotearoa

Peak season
March to June

Size
Cap: 10–15 cm across; Stipe: 5–20 cm long

Amanita muscaria, the fly agaric, is probably the world's most famous fungus. She's incredibly iconic, and doubly striking with her cherry-red cap and bright white spots. This species is introduced to Aotearoa and is easy and delightful to spot. The top observer on the iNaturalist website has over 2000 finds!

When I first started looking for fungi, I was shocked to see my first fly agaric. I'd assumed that anything so visually brilliant must be rare. But no — go walking through some pines in autumn and you can spot not one, but tens, sometimes even hundreds, of these crimson caps.

Dig a bit deeper, and you'll find that this species has a whole heap of stories to tell. Fly agarics contain ibotenic acid and muscimol, two psychoactive substances. Homemakers of old used this to their advantage. They'd drop a few chunks of it into a cup of milk and wait. Pesky houseflies would then drink the milk, become disoriented and fly into walls, neutralising themselves.[12]

Amanita muscaria may also be partly responsible for our visions of Santa Claus and his flying reindeer. Ancient Nordic shamans were known to deliver red and white magic mushrooms to people during the winter. They usually entered snow-blocked homes through the chimney and used reindeer, who love to eat fly agarics, to get around.[13] The colour scheme matches, too. Red and white *Amanita* grow under pines, and that's where we place our gifts wrapped in paper of the same tones.

One of the fastest ways to get people in a fungi identification Facebook group into an argument is to ask if this species is 'eatable'. There's lots of evidence that ancient cultures used this powerful fungus for mystical experiences, but there's no doubt that it contains powerful toxins, too. Either way, I find it pretty wild that something so magic hides in plain sight.

Juvenile fly agaric.

Death cap

Scientific name
Amanita phalloides

Distribution
Has been found in the Auckland, Bay of Plenty, Waikato regions in the NI and in the Nelson and Tasman regions of the SI. This deadly species may grow in other areas too

Peak season
March to July

Size
Cap: 5–15 cm across; Stipe: 8–15 cm long

This fungus's common name says it all. It will kill you. Or try its very best to. But only if you eat it, of course. Each death cap contains enough toxin to kill an adult human. And they're responsible for about 95% of mushroom-related deaths worldwide.

This is a species every forager should be aware of, especially if you're foraging for field mushrooms or other gilled varieties. The death cap grows under introduced trees, especially oaks, has an olive-green cap, sometimes with a tan/yellow sheen, white gills, a white stipe and spores, a skirt, and it grows from a defined volva. The volva especially sets it apart from edible species like field mushrooms. Death caps reinforce the golden rule of fungi foraging, *when in doubt throw it out*.

The death cap's toxins are called amatoxins. They cannot be 'cooked out' and cause acute cell death, basically causing your liver and then your other organs to liquefy. What makes this species even scarier is the lag time — it can take up to three days for symptoms (vomiting and diarrhoea) to show up. This makes it harder to find the link and get the right treatment in time.

Socked flycap

Scientific name
Amanita pekeoides

Distribution
Found throughout the NI apart from the Taranaki region and throughout the SI apart from Central Otago

Peak season
October to July

Size
Cap: up to 8 cm across; stands up to 11 cm tall

This endemic species has unique striations, or stripes, that run from the olivey brown centre of its cap to the edge. This species also has gray warts on its cap. The stipe has a dark, rough texture. Like all *Amanita* species, this one has white spores. Look for it in beech forests and among mānuka and kānuka in autumn.

Jewelled amanita

This introduced species looks a lot like the fly agaric, but its cap is creamy yellow. As with the fly agaric it has white gills, a white stipe, and grows from a volva. It has large felty white scales on its cap. Look for it growing beneath introduced trees like pines.

Scientific name
Amanita junquillea

Distribution
Found in the NI south of Taupō

Peak season
March to November

Size
Cap: 3–8 cm across

Southern beech amanita

This *Amanita* species is found only in Aotearoa. Its cap is dark brown, sometimes with lighter patches, and blue undertones. The cap has soft grey-brown scales. Its gills are white, turning slightly yellow with age. The spores are white. Look for it growing beneath southern beech trees, as well as mānuka and kānuka.

Scientific name
Amanita nothofagi

Distribution
Found across both the NI and SI, apart from South Canterbury and Central Otago

Peak season
November to July

Size
Cap: 3–13 cm across

Honey mushrooms

Honey mushrooms all fall into the *Armillaria* genus, which is also home to the biggest living thing on earth — a humongous fungus that lives in Oregon's Malheur National Forest in the United States. It's a network of honey mushrooms and is believed to be the largest single living thing on Earth (by biomass). It covers around 965 hectares, which is about 1665 rugby fields or roughly 10 square kilometres. It's thought to be around 2400 years old.[14]

While we don't have any honey mushroom blooms quite so large here, we do have some pretty cool varieties. One species may have been used by early Māori as a food resource, and there's work being done now to see if this fungus could help us slow the spread of pesky wilding pines (see page 42 for more on that). Look for honey mushrooms on dead wood, distressed trees and old stumps. They're large, grow in cuddly clusters, and have tough stipes, skirts and beautiful gills; and one species even glows in the dark.

Lemon honeycaps.

Harore

Scientific name
Armillaria novae-zelandiae

Other common names
Olive honeycap, bootlace mushroom

Distribution
Found throughout Aotearoa

Peak season
March to July

Size
Cap: about 6 cm across; stands about 10 cm tall

This honey mushroom usually grows in large groups on dead wood. The young ones that haven't yet opened have small warts on their caps, called squamules. As the caps open, the squamules fall away and sometimes completely disappear, leaving behind a smooth, slightly sticky, olive-coloured cap.

This species may have been gathered and eaten by Māori, and it is an aggressive decomposer.[15] It can infect both native and introduced trees, and is sometimes called the bootlace fungus because its mycelial strands are sometimes just that thick.

Look for this species in both native bush and among introduced trees; anywhere there's plenty of dead wood. If you spot some, return at night, give your eyes plenty of time to adjust to the darkness, and you might be able to see their bioluminescent glow.

Lemon honeycap

Scientific name
Armillaria limonea

Distribution
Grows throughout the NI and most of the upper SI north of Christchurch. Some sightings in Southland and around Dunedin

Peak season
April to July

Size
Cap: about 6 cm across; stands about 11 cm tall

This honey mushroom has a pastel lemon colour to it and little brown scales on its cap. It usually grows in a cluster, with many individuals emerging from one central point. Look for these on fallen logs and old stumps in the cooler months, on wood from both native and introduced trees.

Cortinarius vinicolor.

Webcaps

Webcap species are in the *Cortinarius* genus and share a unique feature that's noted in both their common and scientific names — many have a cobwebby material covering their gills when they first emerge. This is a partial veil, which is sometimes called a cortina. The thready tatters of the cortina can sometimes be seen hanging from the cap or stipe of *Cortinarius* mushrooms once they've opened.

There are heaps of *Cortinarius* species here in Aotearoa. Over 150 species have been described here, and their colour palette knows no bounds. You'll find them in midnight purple, sea-green, glistening yellow and more. Not only do these fungi add pops of colour to our forests, but some showcase evolution for us in real time.

Gilled corts

Scientific names
Too many to list; the stars are featured on pages 224–27

Common names
Corts, webcaps

Distribution
Found across Aotearoa

Peak season
March to September

There are just so many gilled cort species here in Aotearoa. Some are blue, others are pink or even golden. Some are slimy, others are drier. All feature the same web or cortina that covers the gills before they open.

To spot a wide variety of corts, check under *Nothofagus* beech trees and mānuka trees, which they grow in association with. If you really want to dig into your cort identification, always note down what trees you find nearby. Cross-reference your finds with the cort section in *A Field Guide to New Zealand Fungi* by Shirley Kerr. This book features pages and pages of their brilliance.

While I've only included a few of the most notable gilled cort species here, I cannot overstate just how varied this group is. It's one of the only groupings in this book where I have truly seen a species in every colour of the rainbow. Corts are constantly putting on a show for us. When you can get your eyes on and notice them, it's pretty spectacular.

A gilled *Cortinarius* species with tiny mites on it.

Cortinarius aerugineo-conicus

Common names
None yet

Distribution
Manawatū, Wellington, Tasman, West Coast, Canterbury and Southland

Peak season
April to November

Size
Cap: up to 50 mm across

This species has only been spotted a few times, but it's unique to Aotearoa and has a special colour. Its cap has a tie-dye pattern featuring a cool ice-blue and creamy white. Check under beech trees for this whimsical cort.

Canary webcap

Scientific name
Cortinarius canarius

Distribution
Auckland, Waikato, Manawatū and Wellington in the NI, and throughout the SI apart from South Canterbury and Otago

Peak season
March to June

Size
Cap: up to 75 mm across; stands up to 90 mm tall

A common yet colourful cort that's canary yellow from top to bottom. Quite large, which makes it even easier to spot. This native species lives beneath beech trees.

Bearskin webcap

Scientific name
Cortinarius ursus

Distribution
Waikato and Wellington regions of the NI, and Nelson, Tasman, West Coast, Canterbury and Southland in the SI

Peak season
February to July

Size
Cap: up to 100 mm across; stands up to 120 mm tall

This cort species with a shaggy cap is only found in Aotearoa. The shag has a red-brown colour to it and the stipe is grey to brown. Look for this species below beech trees; it's mostly been found in the South Island so far.

Cortinarius tessiae

Common names
None yet

Distribution
Bay of Plenty, Manawatū, Wellington, Tasman, West Coast, Southland and South Otago

Peak season
April to October

Size
Cap: up to 50 mm across; stands up to 60 mm tall

This endemic species looks like all the highlighter colours run together. The cap is a blend of blue-green at the start, but this morphs into orange-yellow with time. It also has a sticky texture, which adds a glossy varnish. The gills are a greyish purple. Find it below beech trees.

Elegant blue webcap

Scientific name
Cortinarius rotundisporus

Distribution
Northland, Auckland, Bay of Plenty, Waikato, Hawke's Bay and Wellington regions in the NI. Nelson, Tasman, West Coast, Canterbury, Southland and Otago in the SI

Peak season
April to July

Size
Cap: up to 50 mm across; stands up to 75 mm tall

These caps almost look like a cosmic night sky. Blue-green morphs into the greyish centre, and the light lilac-coloured gills are similarly beautiful. This native species is found under mānuka trees.

Cardinal webcap

Scientific name
Cortinarius cardinalis

Distribution
Found in the NI south of Taupō and across the SI apart from Marlborough, South and Mid Canterbury, and Otago

Peak season
March to July

Size
Cap: up to 60 mm across; stands up to 90 mm tall

This webcap has a brilliant red colour. Its common name comes from its colour's similarity to the robes of Roman Catholic cardinals. The gills are more orange and it often has a bell-shaped cap. The base of the stipe has a red, fibrous texture.

Cortinarius carneipallidus

Common names
None yet

Distribution
Northland, Bay of Plenty, Hawke's Bay, Manawatū, Wellington, Nelson, Tasman, West Coast, Canterbury and Southland

Peak season
March to July

Size
Cap: up to 100 mm across; stands up to 120 mm tall

This endemic species is midnight-purple. It grows beneath beech trees. Its stipe is fibrous, with strands of deep and pale purple, almost white, running throughout it. The stipe's interior flesh is light in colour. The cap has a speckled texture.

Cortinarius vinicolour

Common name
None yet

Distribution
Found in Auckland, Wellington, Tasman, West Coast of the SI and Southland

Peak season
May to October

Size
Cap: usually 10–25 mm across

This red cort is dry and dark, dark red with a slender, skinny stipe. Its gills are also dark red but become more brown and rusted in colour over time. Look for this striking species in cooler months beneath beech trees.

Secotioid corts

Scientific names
Cortinarius epiphaeus (top) and *Cortinarius* species (bottom)

Common names
Golden pouch (top) and purple pouch (bottom)

Distribution
Hawke's Bay, Manawatū and Wellington in the NI and Nelson, Tasman, Marlborough, West Coast, Canterbury, and around Wānaka and Queenstown in the SI

Peak season
April to June

Size
Cap: up to 60 mm across; stands up to 70 mm tall

Two corts that are completely amazing to spot are the golden pouch and the purple pouch. At first glance they look like classically capped and gilled mushrooms. Turn them over, though, and you'll find that the cap folds in on itself towards the centre. This creates a pouch-like structure around the stipe. The word 'secotioid' describes this unique form where the mushroom's spores are held internally.

Slice one of these fungi lengthways and you'll find that the pouch is brown and fleshy on the inside. This is where the spores develop. While other mushroom spores are forcibly expelled to the environment from their gills, secotioid fungi hold theirs within, protecting them as they mature. And rather than dropping their spores themselves, these fungi rely on animals and insects to have a nibble and spread them that way — on their fur, in their poo, and so on.

Secotioid fungi are sort of like awkward teenagers. They have evolved from classically capped, gilled mushrooms to this pouch-on-a-stem look. They're on their way to becoming fully closed pouch fungi, also known as 'gasteroid' fungi, but are not quite there yet.

Look for these two secotioid cort species in beech forests.

Gasteroid corts

Scientific names
Cortinarius beeverorum (top) and *Cortinarius peraurantiacus* (bottom)

Common names
None yet

Distribution
Auckland, Bay of Plenty, Wellington in the NI; Tasman, West Coast, Southland and around Dunedin in the SI

Peak season
March to October

Size
Up to 25 mm wide by 25 mm tall

Gasteroid corts don't have stipes. They have completed an evolutionary marathon and gone from classically capped mushroom, to enclosed secotioid caps, to a stipe-less gasteroid form.

These red and yellow gasteroid corts can be found in Aotearoa beneath beech and mānuka trees. They look like rubies and yellow sapphires tucked in among the decaying leaf litter. Their interior spore-bearing flesh looks like an Aero choccy bar.

Now, the brilliant colours that corts have aren't just for show. There's a wider story, a clever adaptation, within them, and it involves one of the most storied animals to ever walk this land: the moa. Scientists have examined moa poo fossils, called coprolites, to better understand what moa ate. They found *Cortinarius* species in those poo records and reckon that these fungi may have evolved their bright colours to catch the moa's eye. And it makes sense really — these show-stopping colours sing out from the forest floor and are near-impossible to miss against the deep green background. Which would've been handy for moa, some of which stood about 3 metres high as they foraged about.[16] Moa could've also potentially confused these fungi with podocarp fruits like the similarly coloured miro berry.

Chalkcaps

Russula species, also called chalkcaps, are usually quite dry to the touch. They aren't moist like the waxcaps and they don't put out any liquid like the milkcaps do. Chalkcaps have a crumbly, brittle texture and can be snapped in half like a piece of primary school chalk.

Snails and slugs love to nibble on *Russula* species, which often have a few bites missing. The caps flatten out over time, and often form a dimple in the centre with age. *Russula* gills are creamy white, while the cap colour varies considerably even across the same species.

Look for *Russula* growing from the ground in association with beech and mānuka trees. *Russula* species are ectomycorrhizal with these trees, meaning their mycelium grows through the tree roots to facilitate the exchange of tree sugars for fungally foraged soil nutrients. And keep a keen eye out: sometimes you can spot these wee caps pushing through the forest floor, unfolding themselves and their colour in slow motion.

A purple chalkcap.

Russula kermesina

This species has a unique shape. Its cheerful cherry-red cap rounds out, goes over the edge, and gathers in an orb-like shape around the stipe, often hiding the gills. Its stipe is red at the base and fades to pale yellow towards the top. Slugs and bugs often snack on the bright cap, revealing the white interior flesh and gills. These animals then spread the spores in their own special ways — usually through their poo.[17]

Common names
None yet

Distribution
Bay of Plenty, Manawatū and Wellington regions in the NI. Nelson, Tasman, West Coast, Marlborough, North Canterbury and Southland in the SI

Peak season
February to June

Size
Cap: up to 50 mm across; stands up to 60 mm tall

Camembert brittlegill

Introduced to Aotearoa, this species is often found among introduced trees. I'm unclear on how it got its cheesy common name, but it does have a funky scent. It has a pinky grey to yellowish-brown cap. Look for it growing among introduced trees.

Scientific name
Russula amoenolens

Distribution
Found across the NI apart from Hawke's Bay, and in Nelson, Tasman, Canterbury and around Dunedin in the SI

Peak season
December to June

Size
Cap: up to 60 mm across; stands up to 80 mm tall

Pink oyster mushrooms.

Oysters

Oyster mushrooms are delicious to some organisms and deadly to others. These fleshy fungi grow from dead wood and are attached to that wood on one side by a very short stalk. They have gills underneath which fan out from the point of attachment, and they are usually somewhat soft and flexible. They have white spores. Look for New Zealand oyster mushrooms on dead wood in native ngāhere.

While more taxonomic work is needed to understand our wild native oysters and whether they're edible, cultivated oyster mushrooms are a delight. They come in grey, bright yellow, bubble-gum-pink and even blue. They have a slight seafood flavour — they're also sometimes called abalone mushrooms — and go great in stir-fries. Lots of gourmet mushroom growers in Aotearoa sell oyster mushrooms at farmers' markets.

Now, if you're a nematode (microscopic worm), oyster mushrooms are bad news. These fungi have evolved the ability to hunt, essentially. They construct sticky traps within their mycelium, which ensnare nematodes and allow the fungus to attack them and absorb them as food. Yikes.

Olive oyster

Scientific name
Pleurotus purpureo-olivaceus

Distribution
Waikato, Manawatū and Wellington regions of the NI, and Nelson, Tasman, Canterbury, West Coast and Southland regions of the SI

Peak season
April to August

Size
Cap: 10–70 mm across; stands up to 60 mm tall

The olive oyster can be found on southern beech wood in the cooler months. They have a fan shape and are slightly shiny when wet. Their caps have a unique olive-green colour, but can vary considerably in shade. Their gills are pale grey and they tend to grow in the interior areas of Aotearoa in the mountains and within beech forests.

We aren't 100% sure about the olive oyster's edibility, but it is related to commercially grown edible oyster species. Work is underway to better understand what oyster species we have here and how we might be able to use them as food. At least five other oyster mushroom species grow in Aotearoa.

Panus purpuratus

Common names
None yet

Distribution
Found across the NI apart from Northland. Nelson, Tasman, Kaikōura, Christchurch and Dunedin in the SI

Peak season
October to May

Size
Cap: 80–90 mm across

This purpley fungi is found only in Aotearoa. It isn't technically an oyster mushroom, but its form looks quite similar so I have included it here. This fungus has a unique ochre-purple colour. Its cap is depressed in the centre, giving the fruiting body a distinct funnel shape. It has deeply decurrent gills, which are a light creamy purple. The cap sometimes rolls over the edge giving this species a distinct purple rim. Look for this species on fallen and decaying wood.

Cup fungi

I always find cup fungi extra-magical. Their shape is unique and they often grow in clusters of ten, twenty, or even hundreds in one spot. Some look like tiny birds' nests, some turn our forests blue and others look like spa baths for ants.

Many of these species are quite tiny and require adults to look down their noses or peer over the edges of their glasses to bring them into focus. Michael Kuo, who runs the legendary website MushroomExpert.com, calls this the OMHT — old man head tilt!

Cup fungi hold their spores within the cup structure. They rely on the wind, and sometimes the assistance of raindrops, to spread or splash their spores about.

Cookeina colensoi.

Lemon discos

Scientific name
Calycina citrina

Other common names
Yellow discos, yellow fairy cups

Distribution
Found south of Hamilton in the NI, and across the SI apart from Marlborough and South Canterbury

Peak season
March to June

Size
Discs: 2–5 mm across; they stand just 1–2 mm tall

'Lemon discos' has to be one of my all-time-favourite common names for a fungus. It's so fun and cheerful, which is exactly the vibe that this native fungus exudes.

The discos themselves are tiny, only a few millimetres across. But they live in clusters of tens, sometimes even hundreds, dotting dead logs like bright-yellow polka dots. They have teeny tiny short stipes, but are mostly identifiable by their cheery colour and miniature size. Look for them on fallen twigs and branches in all forest types.

Eyelash cups

Scientific name
Scutellinia species

Distribution
Found throughout Aotearoa

Peak season
September to December

Size
Each cup is 5–15 mm across; they stand up to 3 mm tall

Eyelash cups are fairly common, and fairly glamorous. Their burnt-red cup structures are ringed with curly dark hairs that really do look like eyelashes. Their curl and length are enviable.

Look for these on the side of tramping tracks. They can most often be found growing on very rotten, very mossy wood. Their tiny size makes them almost impossible to photograph without a specialised lens, which makes spotting them extra special — a find only to be enjoyed in the moment. Found in all forest types on damp, squishy, sometimes even partially submerged wood.

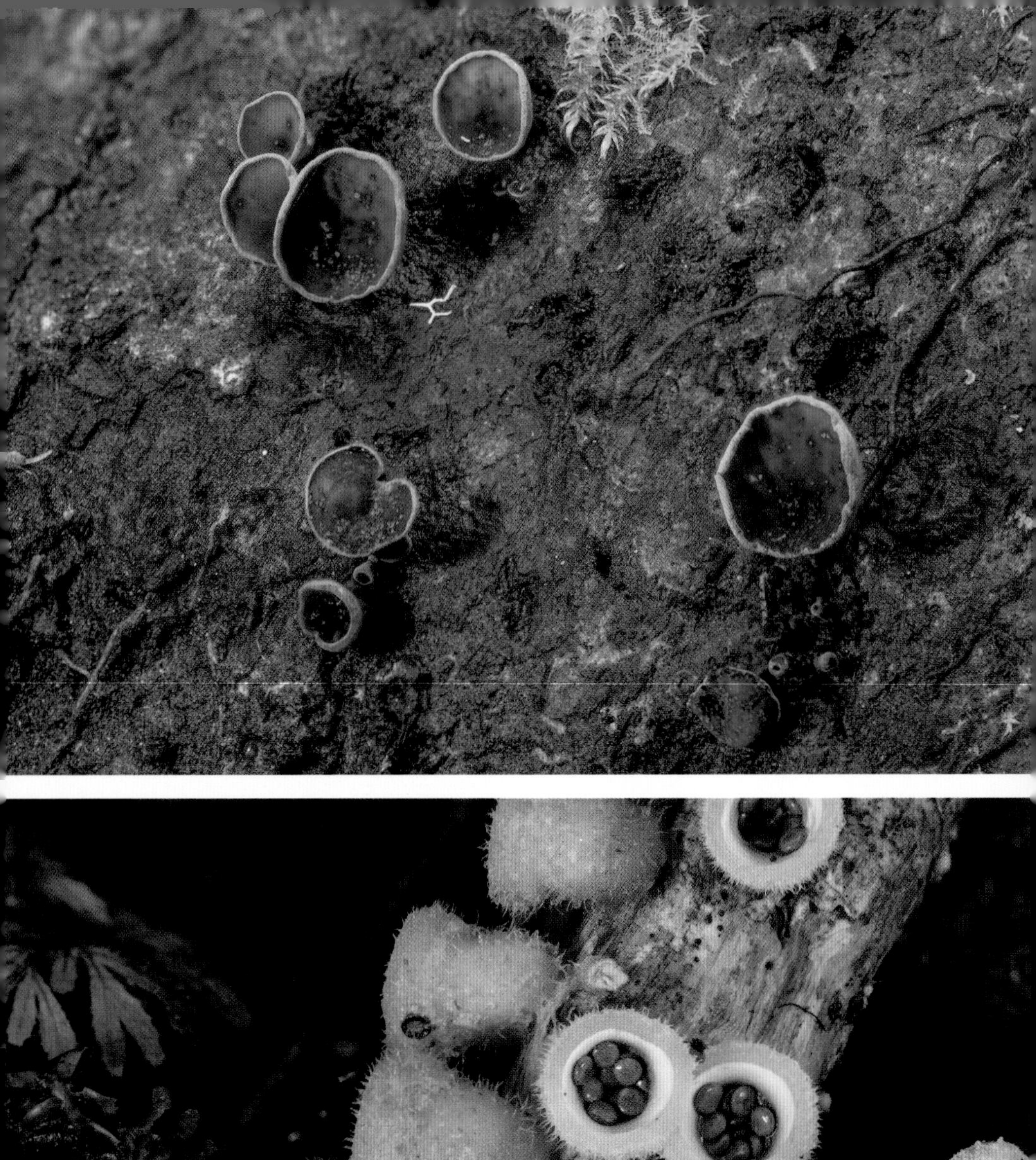

Emerald elf cups

Scientific name
Chlorociboria species

Other common name
Blue elf cups

Distribution
Found throughout Aotearoa

Peak season
The blue stain is always about; the fruiting bodies less so

Size
Each cup is up to 5 mm across; they stand just 1–2 mm tall

Ever seen a bit of blue wood while tramping through the bush? Then you've already seen this fungus, or evidence of it at least.

Emerald elf cups are supremely small, and quite ephemeral. Their bright-blue, cup-shaped fruiting bodies are only out and about sometimes, usually in autumn after a big rain. But their work is almost always on display. They're the watercolourists of the forest, and go about painting/staining wood a brilliant blue.

There are at least 13 emerald elf cup *Chlorociboria* species in Aotearoa.[18] These species produce an organic compound called xylindein, which stains its woody substrate a brilliant blue-green colour. This unique wood, sometimes called blue oak or green oak, is something of a commodity for woodworkers. Its use goes all the way back to fifteenth-century Italy where it was used to make engraved wall panels.

Bird's nest fungi

Scientific name
There are several species within several genera in Aotearoa

Distribution
Found throughout Aotearoa

Peak season
Always around

Size
Each nest is up to 10 mm across; they stand up to 20 mm tall

Bird's nest fungi species are true miniatures. They look like near-replicas of the homes some birds create for their tiny babies.

The 'eggs' within these nests are precious parcels, too. They're called peridioles, and they are packets of spores. When a raindrop connects with the nest, it splashes the spore packets out and launches them into their next phase of life. The nests are often covered when they first emerge, to protect the spore packets while they mature.

Bird's nest fungi can be found on twigs, rotting logs, and in mulched garden beds. There are several bird's nest species within several genera in Aotearoa, and they require a sharp eye to spot.

Orange peel fungus

Scientific name
Aleuria aurantia

Distribution
Found throughout Aotearoa

Peak season
April to September

Size
Each peel is up to 30 mm across; they stand up to 15 mm tall

This orange species is easy to spot on disturbed soil. Check for it alongside tramping tracks, road cuts, and even in your own garden. It grows sometimes as single peels or as bunches of them. When clustered together, they almost look like an orange camellia flower.

Cookeina colensoi

Common names
None yet

Distribution
Found across the NI, and in Marlborough, Canterbury and the West Coast in the SI

Peak season
November to July

Size
Each cup is up to 15 mm across; they stand up to 5 mm tall

These pale-pink cups can be found on dead wood after rain in the cooler months. They have short stubby stipes, and can become more pink with age.

Purple jellydisc

Scientific name
Ascocoryne sarcoides group

Distribution
Found south of Hamilton in the NI and in Nelson, Tasman, West Coast, Canterbury and coastal Otago in the SI

Peak season
Always around

Size
Each disc is up to 15 mm across; they stand up to 10 mm tall

When this purple species is in its sexual stage — ready to reproduce — it has a cup or disc shape. The rest of the time, it's lumpier and jelly-like. Look for this one on dead wood that is very wet.

Brown forest cup

Scientific name
Urnula campylospora

Distribution
Found from Northland to Tauranga, plus Manawatū, Wellington, Nelson, Tasman, West Coast, Canterbury and Otago

Peak season
April to October

Size
Cup: 30–55 mm across; stands up to 10 mm high

This species is big — each cup can be up to 30 mm across. The cups have a slightly downy, black exterior, and a smooth, shiny, leathery, brown interior. They flatten with age. Look for these on dead wood and fallen mānuka branches.

A hedgehog jelly species.

Jelly fungi

Jelly fungi are fun to pat. They're jiggly and soft, yet also incredibly hardy. Their unique cell structures allow them to expand and contract according to moisture levels in their environment. Many jelly fungi can survive for months on end without rain. In times of stress, they shrivel up and collapse down into themselves — protecting, retreating, waiting. And when the rain returns? Bam! They re-emerge, doubling or tripling in size, often in a matter of hours.

Jelly fungi ebb and flow through life. They go quiet and tough for a bit, and then explode with colour and joy once again. One jelly fungus here is also quite yummy. It's called hakeke, or woodear, and was actually a cash crop in the Taranaki region in the late 1800s. You can read about that one in the section on edible fungi, on page 104.

Jelly babies

Scientific name
Leotia lubrica group

Other common names
Slippery lizard tuft, ochre jelly club

Distribution
Found across the NI, and across the SI apart from South Canterbury and North Otago

Peak season
March to August

Size
The head is 15–25 mm across; stands about 60 mm high

Jelly babies feature an olive green 'head' which is irregularly shaped and folds inwards towards the stipe. The stipe is usually a lighter yellow-green colour and sometimes has little green warts on it. Jelly babies normally grow among moist, cushiony moss, either in clusters or on their own. Only a keen fungi-hunter can eagle-eye this find, as their green nature is almost at one with the moss and ferns. Look for jelly babies along the side of tramping tracks nestled in moss, or sheltered beneath rārahu (bracken fern).

Orange button jelly

Scientific name
Heterotextus species

Distribution
Found south of Auckland in the NI and across the SI

Peak season
April to September

Size
Each button is up to 10 mm across and 6 mm tall

Orange button jellies aren't edible, but they look as if they should be — yummy and gummy, like orange-flavoured lollies full of sweet-and-sour flavour. Look for these bell-shaped species on fallen twigs and branches. They love rotting wood and can be found in many forest types, but especially in super-damp ones. I've never not seen orange button jellies in Arthur's Pass in the wintertime.

Beech strawberry

Scientific name
Cyttaria gunnii group

Other common names
Beech orange, myrtle orange

Distribution
Bay of Plenty, Hawke's Bay, Manawatū and Wellington regions in the NI. Found across the SI apart from Canterbury and Otago

Peak season
October to January

Size
Fruiting body can be up to 30 mm by 30 mm

Beech strawberries look like odd orange golf balls. They grow on *Nothofagus* beech trees and often fall to the forest floor. They're only found in Aotearoa, Australia, Chile and Argentina, so they're a living reminder of deep time . . . a time when those countries were all part of one supercontinent, Gondwanaland. 180 million years ago, Gondwanaland broke up and the modern continents began drifting apart. Flora, fauna and fungi with modern distribution patterns like that of the beech strawberry have helped us unravel what the world looked like back then.

Beech strawberries are gathered for food in some cultures. The Araucans of Chile actually used it to make alcoholic bevvies. Beech strawberries are coated in the same kind of yeast used in classic beers like Speight's.[19]

Darwin found a similar species on his forays in South America. He brought one back to Kew Gardens in England, but the sample got lost. Some 180 years later it was found in a dusty archival basement, still intact, perfectly preserved in a pickle jar.[20] Nice one, Charles.

Snow fungus

Scientific names
Tremella fuciformis

Distribution
Found across Aotearoa apart from Taranaki

Peak season
Always around

Size
Up to 70 mm across by 40 mm tall

Look for snow fungus in bark crevices or on dead branches after rain. It is translucent, almost shimmery, and very jiggly. It shrivels up to almost nothing when there is no rain. A similar species, *Tremella mesenterica*, is a bright yellow-orange. It's called witches' butter and has a brain-like texture — perfect for a storybook witch to spread on toast. Both species can shrink and expand in accordance with moisture levels many times over the course of their lives.

Hedgehog jellies

Scientific name
Pseudohydnum species

Distribution
Auckland, Waikato, Bay of Plenty, Manawatū and Wellington regions of the NI, and in Marlborough, West Coast, Southland and South Otago in the SI

Peak season
April to July

Size
Up to 40 mm across

Hedgehog jellies range from pale brown, to pale gray, to white and even near-translucent. Their most defining characteristic though is their texture. Their undersides are toothy, like a hedgehog. These species are also a bit sparkly and grow on dead wood or woody debris. We have at least two native hedgehog jelly species in Aotearoa.

Stinkhorns

The stinkhorns take the cake when it comes to being strange, alien, semi-gross and hard to believe. They have major underworld vibes and come in all sorts of shapes — from basket-like to super-phallic. This group is bizarre.

One key feature that the stinkhorns share is their spore surface. Classic, capped gill mushrooms drop tiny puffs of spores into the wind, in a gentle and almost secretive way. Stinkhorns, on the other hand, wear their spores on their sleeves. Their fruiting bodies generally have smears of poo-brown goo on them. This sticky substance contains their reproductive spores, and it's smelly. The smell attracts flies and other insects, who inadvertently spread the fungal spores far and wide. Clever on the fungus's part, really.

Golden basket stinkhorn.

Tūtae kēhua

Scientific name
Ileodictyon cibarium

Other common names
Tūtae whatitiri, tūtae whetū, paru whatitiri, whareatua, white basket fungus

Distribution
Found across Aotearoa

Peak season
April to August

Size
Up to 200–250 mm across

I'd seen photos of this native fungus online and assumed that it only grew in the deepest, densest parts of the bush. But no. I spotted my first-ever specimen outside the Auckland international airport terminal in a mulched garden bed. Despite being completely other-worldly, you can often find this species in the most ordinary places — I've also spotted them in a Four Square car park and next to the Margaret Mahy playground in Christchurch's CBD.

There are over 35 Māori names recorded for this fungus. Many of them include the word whatitiri, meaning 'thunder', which may refer to its emergence after big storms. The name tūtae kēhua offers another explanation: ghost droppings. Early Māori may have used this species for food, gathering them before they opened and cooking the outer layer of flesh.[21]

This species' 3-D structure is unmistakable, and so is their mobility. When this fungus first emerges it looks almost like a puffball. After a bit of time, though, the outer layer peels away and the basket structure springs forth. These baskets occasionally catch the wind and roll away, spreading their spores to new territories, like a fungal tumbleweed or zombie soccer ball.

Some Cantabrians will remember that the playground in the botanical garden used to feature a piece of tūtae kēhua-shaped climbing equipment. Keep an eye out for it next time you're in Lyttelton. It lives on.

Puapua-a-Autahi

Scientific name
Aseroe rubra

Other common names
Anemone stinkhorn, starfish fungus

Distribution
Found across Aotearoa

Peak season
October to July

Size
Up to 80 mm across; stands 60–100 mm tall

This native species is truly wild. Like a red squid reaching out from below, this scarlet fungus coated in shiny, slimy brown goo shocked me the first time I saw it. But I didn't clock it with my eyes first. My partner and I were walking along the Queenstown waterfront and had just entered an area of pines. Something smelled horrible, but there were lots of *Amanita* about so I was happy and fungi-spotting as we went along. And that's when I saw it. A red octopus of the land.

Puapua-a-Autahi, the anemone stinkhorn, is striking and spooky all at once. Its smell is arresting and forces you to pay attention. It grows from a girthy stalk. When it first emerges, the tentacles are often gathered together in the centre, like a chef's kiss. But with time, they open wide and unfurl like the wavy arms of an anemone, a starfish, or some other coral reef resident.

In the centre of the star-like shape you'll find the glistening brown goo, the substance that contains the spores. This is the smelly stuff. Its rotting-flesh scent attracts spore-spreading flies as well as other interested passers-by. Like me. This fungus just doesn't seem real; but it is!

There are lots of stories in the Māori name for this fungus. Autahi is also known as Canopus, the second-brightest star in the sky, and 'puapua' has many definitions: wreath, petal, shield.[22] I can see them all in this species.

Stinky squid

Scientific name
Pseudocolus fusiformis

Distribution
Northland, Waikato, Bay of Plenty, Taranaki and Manawatū regions of the NI. Tasman, West Coast and Southland regions of the SI

Peak season
November to May

Size
Up to 25 mm across and 120 mm tall

This fungus looks like a spooky pink-orange hand coming up from the ground. The fingers emerge from a greyish volva and have dabs of brown spore-bearing goo on their interior sides. Look for this one in native bush and in mulchy gardens.

Red dog stinkhorn

Scientific name
Mutinus ravenelii

Distribution
Has been spotted across the NI, predominantly north of Taranaki, with one or two sightings around Wellington

Peak season
November to January

Size
Up to 8 mm across; stands up to 100 mm tall

This elongated stinkhorn emerges from a volva, has a white to red body with a foam-like texture, and tapers to a point at the top, which is coated in spore-bearing brownish-green goo. Flies love this one. Look for it in parks, gardens, mowed grass and mulchy garden beds.

Golden basket stinkhorn

Scientific name
Clathrus chrysomycelinus

Distribution
Has been seen in Northland, Waikato and Bay of Plenty

Peak season
November to June

Size
Up to 60 mm across by 100 mm tall

Very similar to tūtae kēhua (see page 259), but this one stays attached to the volva from which it emerges. Its spores are held in gooey droplets within the basket. The basket itself is white towards the base, and yellow-red towards the top. Look for it on the ground in broadleaf podocarp forests after autumn rains.

Devil's fingers

Scientific name
Clathrus archeri

Distribution
Found throughout Aotearoa

Peak season
October to June

Size
Each arm can grow up to about 100 mm

This fungi has 4 to 6 red-orange fingers/arms on average. Also known as the octopus stinkhorn, it can be found across the country in leaf litter and mulch beds in areas with high moisture levels. Its stinky spore-bearing goo smells like rotting flesh. *Shudder.*

Pagoda fungus.

Shelves and brackets

The shelves and brackets in this section share a common shape, and all grow from wood. Tonnes of species and genera could fit the bill, but the ones included here are some of the most common, useful and interesting ones.

These fungi all stick out like shelves, but they vary considerably in size and texture. Some are huge, hard and woody. Others are small, soft and squishy. There are even some leathery ones and some that can live for years and years. A few of the species within this group are quite useful, medicinal even, whereas others are invasive and pest-like. One has helped humans 'carry' fire for centuries.

Shelves and brackets can be found pretty much anywhere, so let's start with some of the ones you're most likely to spot in Aotearoa.

Turkey tail

Scientific name
Trametes versicolor

Distribution
Found across Aotearoa

Peak season
Always about

Size
Fruiting bodies are up to 8 cm wide by 5 cm tall

Turkey tails can almost always be found. They often grow from branches and twigs in urban areas. They are slightly hairy, have a fan shape with concentric rings of colour, and have a creamy white underside. These layers of colour vary hugely, and no two are the same. I've seen turkey tail in neutral tones from buff to chestnut, and another bloom that ranged from baby blue to navy. Turkey tail often starts to get a bit green as it ages; this is usually caused by algae making themselves at home on top of it. An entire community on one turkey tail.

Turkey tail is also one of the most well-known 'functional fungi' species. Functional fungi species are seen as offering health and wellness benefits to us humans. For example, lion's mane (*Hericium erinaceus*) helps with mental clarity and sharpness, and *Cordyceps* boosts energy. Turkey tail is a potentially powerful cancer-fighting fungus. It has a high concentration of polysaccharide K, which has been used to treat cancer, and has shown promising results in scientific studies.[23]

Artist's brackets

Scientific name
Ganoderma species

Distribution
Found across Aotearoa

Peak season
Always about

Size
Variable – some species can grow up to 40 cm or more across

Woody and bracket-shaped *Ganoderma* species aren't super-flashy, but they are super-common. You may have already seen some on trees or fallen logs. They're around all year, and have a shelf shape, a grey to brown top and a bright white underside. The underside can be easily bruised, or drawn on — hence the common name. These brackets are also massive: up to 600 mm across and 100 mm tall. Watch one long enough and you might spot it releasing a tiny cloud of spores to the world.

While not found in Aotearoa, *Ganoderma sichuanense*, also called reishi, is the most well-known species within the genus and has been recognised in China as a medicinal mushroom for over 2000 years. It's believed to improve memory, slow ageing, enhance energy and is recommended for insomnia, dizziness and shortness of breath.[24] *Ganoderma* first began appearing in Taoist art in about 1400 AD, around the same time that the first Polynesian peoples may have arrived in Aotearoa.[25]

Ganoderma specimens can live for years and years and years. Watching them expand slowly, morph, and grow over time is spectacular. If you spot one near home, keep checking in on it.

Pūtawa

Scientific name
Laetiporus portentosus

Other common names
Pangu, punk, bracket fungus, morepork bread

Distribution
Found across the NI and in the SI north of Christchurch

Peak season
Always around

Size
Up to 10–30 cm across and 6 cm thick

This bracket fungi isn't the most eye-catching, but it holds old and fascinating stories. Pūtawa grows from beech trees; it has a fawn-coloured top, with a lighter underside that has visible pores. The depth of the pores can sometimes be seen from the side (as seen on the opposite page).

In an often-repeated story, early Māori may have used pūtawa to carry fire from place to place. It burns very slowly, can hold an ember for several days, and can only be extinguished by total smothering. This would have been super-handy when trying to start a fire quickly — something that can be at times near-impossible in the damp native ngāhere. Colonial settlers may have even learned from Māori that this fungus could be used to light smokes when touch papers ran out. Similar species in the northern hemisphere have long been used to start fires and were even sent to soldiers in World Wars I and II to help them light their pipes.[26]

Pūtawa also had medical applications. It was used to treat wounds and sores, and to keep pressure on them when needed. Patches were sometimes made out of this fungus by cutting it into strips and tying it around wounds. Its absorbent, elastic quality made it a useful natural bandage.[27]

Pagoda fungus

Scientific name
Podoserpula pusio

Distribution
Found across the NI apart from Taranaki. Also across the SI apart from Marlborough, South Canterbury and North and Central Otago

Peak season
April to August

Size
Up to 3 cm across; stands up to 10 cm tall

I've never seen this one myself, but I dream of the day. Pagoda fungus grows upon itself, like a bookshelf or tiered wedding cake. Its colour ranges from white to light pink, so the colour scheme feels on-brand for that too. I'm not aware of any other fungi that grow like this. Its common name, pagoda fungus, makes me think of a beautiful zen garden for tiny forest-dwellers. Insects hang out on every level, according to size.

Look for this species among rotten wood in all forest types, and pine stands in particular. Sometimes it even grows within rotten stumps. How magic.

Orange ping pong bat

Scientific name
Favolaschia claudopus

Other common name
Orange pore fungus

Distribution
Found across the NI; along the coast and in coastal foothills north of Christchurch as well as on the Otago Peninsula in the SI

Peak season
February to July

Size
Fruiting bodies can be up to 2 cm across and 4 cm tall

This introduced and unmistakable species is high-vis. It looks like perfectly sized ping pong bats for birds, in tradie-orange.

People who experience trypophobia, a fear of holes, do not get on well with this species. But others, like me, find its repetition incredibly satisfying to gaze at. One thing I find really cool about this species is the size variation. On one single dead twig you might spot specimens the size of your thumbnail all the way down to tiny ones the size of a pin.

This traffic-cone-coloured fungi can be found on wood of all kinds, and sometimes grows like a weed in the native bush where it is invasive. Look for it after autumn rains.

Flabby poreconch

'Flabby poreconch' feels like an undercooked way to describe this native species, which grows from tawa trees. It's pearlescent and semi-translucent, able to catch the light and redirect it through its perfectly formed matrix — a fungal version of the prisms you hang in the kitchen window to cast a rainbow on the floor. This pattern, which creates a bumpy feel on the cap, and the rubbery texture will help you identify this species.

Scientific name
Favolaschia pustulosa

Other common name
White porecap

Distribution
Found across the NI except Hawke's Bay and Gisborne. Nelson, Tasman, West Coast, Otago and Southland in the SI

Peak season
November to July

Size
Fruiting bodies can be up to 2 cm across and 5 cm tall

Sunset leather bracket

This fairly common native species looks like a little fungal sunset. It has layered rings of deep yellow, orange and red on top, and a smooth underside. As the fungus grows, these lobed arcs of colour sometimes meet up and form a cup. It's slightly velvety and dry to the touch. This species grows from branches and logs in all forest types.

Scientific name
Stereum versicolor

Distribution
Found across Aotearoa

Peak season
Always around

Size
Fruiting bodies are 6–9 cm across and 5–12 cm tall

Pouch fungi

Pouch fungi don't have pores or gills. Instead, they have fully enclosed caps that look like pouches. These little fungal purses hold all their spores within. They rely on animals to spread their spores, rather than dropping them themselves.

The pouches are often lumpy and bumpy. Occasionally they gobble the stipe up almost entirely, so you may have to clear away the leaves and dirt to spot it. We have quite a few interesting pouch species here. They come in red, blue, purple and gold, and can usually be found among the leaf litter, like half-buried gems.

Blue pouch fungus.

Psilocybe weraroa

Common name
None yet

Distribution
Found across the NI; has been spotted in Tasman and Southland in the SI

Peak season
April to August

Size
Pouch: up to 4 cm across; stands up to 3 cm tall

One of the most distinct pouch fungi found here is *Psilocybe weraroa*. It has a pale-green to grey-blue colour, a whitish stipe, and contains psilocybin, a psychoactive substance that can induce hallucinations. In short, this is an endemic magic mushroom.

This species grows from dead wood that is often partially buried in the soil, and is mostly found in the North Island in all forest types. Look for it from April to August, after warm rains but before the slugs make a meal of them.

Work is underway to determine just how much psilocybin this species contains, and to investigate its potential to be used within a Māori cultural framework to treat addiction. You can read more about that on pages 47–50. Research conducted in Aotearoa and abroad in recent years has signaled that psychedelics, including magic mushrooms, may have potentially positive applications in the treatment of addiction, depression and other mental illnesses. Foraging magic mushrooms outside of research like this is inadvisable and also illegal.

Scarlet pouch

Scientific name
Leratiomyces erythrocephalus

Distribution
Found across Aotearoa apart from Central Otago

Peak season
April to October

Size
Pouch: up to 3.5 cm across; stands up to 5 cm tall

This native species' cherry-red cap and yellowish stipe are unmistakable. Look for it on rotting wood or in soil that has woody debris throughout it. It is usually found in native bush and mulched gardens, and is sometimes seen beneath miro trees, which have berries of a similar colour.

Blue pouch

Scientific name
Clavogaster virescens

Distribution
Found across Aotearoa apart from Canterbury and North and Central Otago

Peak season
March to September

Size
Pouch: up to 3 cm across and 6 cm tall

This pouch is a bit more elongated and pointed than some of the other pouch fungi and it's sometimes called the spindle pouch. It has a large blue cap and whitish yellow stipe, which is sometimes partially hidden by the soil and leaf litter (as seen below). Only found in Aotearoa, it grows from dead wood in all sorts of forests.

Golden pouch

Scientific name
Cortinarius epiphaeus

Distribution
Found in Hawke's Bay, Manawatū and Wellington in the NI and Nelson, Tasman, Marlborough, West Coast, Canterbury, and around Wānaka and Queenstown in the SI

Peak season
April to June

Size
Pouch: up to 6 cm across; stands 4–8 cm tall

This *Cortinarius* species has no gills, and is only found in Aotearoa. It has a solid yellow colour to it, with a brown undertone. The stipe is usually lighter in colour. Look for it in autumn in southern beech forests.

Purple pouch

Scientific name
Cortinarius species

Other common name
King's pouch

Distribution
Hawke's Bay, Manawatū, Wellington, Nelson, Tasman, Marlborough, West Coast, Canterbury, Wānaka and Queenstown

Peak season
April to June

Size
Pouch: up to 7 cm across; stands up to 7 cm tall

There are several purple pouch species within genus *Cortinarius*. These do not have gills but they do have striking purple caps and whitish stipes. They are found under southern beech trees in the cooler months.

An earthstar.

Puffballs, earthballs and 'truffle' types

The fungi in this group are super-fun to pat. They're stipe-less and potato-like. They grow from the ground and are often nestled in loose topsoil or leaf litter. These fungi can be found in all sorts of environments up and down Aotearoa. They hold their spores inside their orb-shaped bodies, and sometimes spread them via a hole at the top, or as the outer surface ruptures and cracks as it ages.

The giant puffball, which we learned about in the section on edible fungi, can grow up to 50 cm across, weigh several kilograms, and is sometimes called tofu of the woods for its texture and flavour.

Look for puffballs, earthballs and 'truffle' types in both urban and rural areas, and in both native and exotic tree stands.

Earthstars

Scientific name
Geastrum species

Distribution
Found across Aotearoa

Peak season
March to July

Size
Up to 5 cm across by 4 cm tall

There are several earthstar species in Aotearoa; they all look fairly similar and can be found in early autumn. They tend to grow in clusters in the leaf litter, like tiny constellations that have fallen to Earth. Earthstars have a tough outer layer that splits and peels away in a star formation, presenting the inner puffball which contains the spores. Native Americans used earthstars medicinally, and they were sometimes seen as indicators of supernatural events.

Look for our native earthstars in the leaf litter of broadleaf podocarp forests. Introduced earthstars are sometimes found in unique forests like the redwood stands in Rotorua. Sometimes the star shape will curl back so much that it almost functions as slits which hold the inner puffball up to the sky, making it easier for you to spot.

Hotlips puffball

Scientific name
Calostoma rodwayi

Distribution
Bay of Plenty, Hawke's Bay, Manawatū and Wellington regions of the NI. Found across the SI apart from Marlborough and South Canterbury

Peak season
April to October

Size
Up to 2–2.5 cm across; stands up to 4 cm tall

This native dark-brown puffball sits atop a slimy, fibrous golden-brown stalk and presents a bright orange 'kiss' to the sky. The orange lips cover the opening where the spores emerge as the puffball ages. Brownish warts cover the rest of the puffball. When this fungus first emerges, a protective grey layer covers the orange lips and brown warts. Look for this one in autumn, on the ground in southern beech forests.

False truffles

Scientific name
Rhizopogon species

Distribution
Found across Aotearoa

Peak season
February to April

Size
Up to 5 cm across by 3.5 cm tall

False truffles aren't the flashiest fungi, but you are liable to see them in pine stands. They are often half-buried among pine needles; wilding pines may even be using false truffles' mycelium to help them spread. False truffles have a yellowy exterior, a potato-like shape and a whitish interior when young, which turns more black with age. False truffles do not have the iconic aroma of real truffles.

Scarlet berry truffle

Scientific name
Paurocotylis pila

Distribution
Found across Aotearoa

Peak season
December to August

Size
Up to 3.5 cm across

I thought this fungus was a bit of chewed-up gum the first time I saw it. It's smooth but not glossy, round but a bit lumpy, and can be found in parks, gardens and within pine forests. Look for it in autumn after rain. It's easy to confuse with the fruits of podocarp trees like miro and rimu. The interior is fleshy and white, but not completely solid — it has winding, empty chambers.

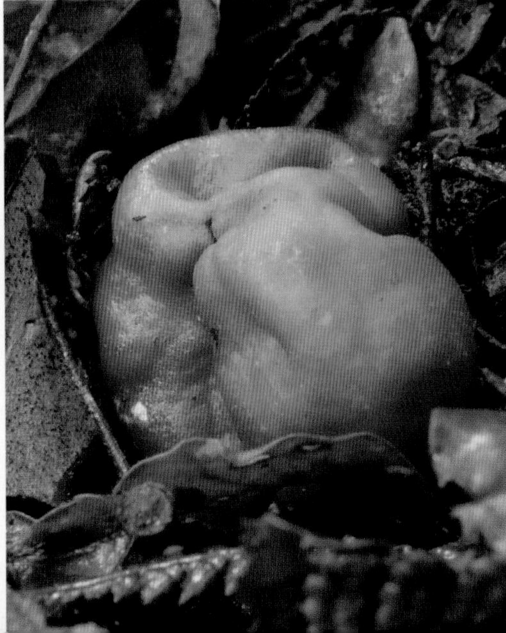

Blue-green potato earthball

Scientific name
Rossbeevera pachyderma

Distribution
Found south of Auckland in the NI, and across the SI apart from South Canterbury

Peak season
December to August

Size
Up to 5 cm across by 4 cm tall

This endemic fungus has a mottled top that stains yellow and a brown interior. It's found beneath southern beech trees and was named in honour of the late mycologist Ross Beever, who added a lot to our understanding of the fungi of Aotearoa.

Violet potato fungus

Scientific name
Gallacea scleroderma

Other common name
Velvet potato fungus

Distribution
Auckland, Manawatū, Wellington and SI apart from South Canterbury and Otago

Peak season
January to August

Size
Up to 10 cm across

This stunning purple species grows in association with beech trees and is only found in Aotearoa. It has a potato-like shape with lumps and grooves, a slightly hairy or scaly texture, and olive-brown interior flesh. Look for it in southern beech forests.

Corals and clubs

Under the sea or under the trees? Coral and club fungi will have you double-checking what biome you're in. They come in all sorts of colours, and are also referred to as clavarioid fungi. These species usually have branched or simple singular fruiting bodies that stand up straight. Most are saprobic decomposers found on soil, decomposing vegetation and dead wood.

There are quite a few coral and club species here in Aotearoa. They are scattered across many different genera. Not all are closely related on the genetic level and microscopic features are usually required to make an accurate identification. Nevertheless, here are some of the most beautiful and unique corals and clubs to find here.

An orange club fungi.

Violet coral fungus

Scientific name
Clavaria zollingeri group

Other common name
Magenta coral

Distribution
Found across the NI and in Nelson, Tasman, West Coast and Dunedin in the SI

Peak season
April to September

Size
Up to 7 cm across and 10 cm tall

Violet coral has a brilliant magenta colour and an unmistakable shape. Look for it below tree ferns in the North Island and upper South Island. This coral has rounded branches that fade from magenta to brown. I have yet to find this species, it's not super common, but it's definitely on my fungi-spotting bucket list. Purple and blue coral fungi just feel otherworldly.

Pterulicium species

Common names
None yet

Distribution
Not extremely well documented but have been sighted in Northland and Bay of Plenty

Peak season
July to September

Size
Up to 4–5 cm long

Pterulicium species have a distinctively spiky quality to them and almost look spiny enough to prick your finger on, like a cactus. From above, they look like sharp corals or fungal porcupines. There are at least three of these spiny species in Aotearoa, and about 50 have been documented globally. One has antibiotic qualities. Look for these unique fungal forms in forests with plenty of rotting wood.

Lilac coral fungus

Scientific name
Ramariopsis pulchella

Distribution
Found across the NI, and across the SI apart from South and Mid Canterbury and Central Otago

Peak season
April to September

Size
Can grow up to 2 cm high

This purple species is native to Aotearoa. Its brilliant hue seems to intensify at the ends of each branching section, giving it a striking glowing quality. Some specimens are paler and more mauve. Look for this unique species growing from damp soil clumps and among decaying leaf litter. Occasionally, it grows from rotting logs.

Artomyces species

Distribution
Auckland, Waikato and Manawatū regions in the NI, and Nelson, Tasman, West Coast and Canterbury in the SI

Peak season
December to June

Size
Varies, but roughly 8–9 cm tall and 4 cm wide

Artomyces species are delicate and have intricate, candelabra-shaped tips. Each branch ends in four or five distinct, finger-like nodes. This gives the appearance that each branch is topped with a tiny crown. This *Artomyces* coral's grey-brown colour isn't the most striking, but its structure surely is. Look for these species in summer and autumn in all forest types on fallen branches and logs. There are at least five *Artomyces* species here in Aotearoa.

Calocera species

Common names
None yet

Distribution
Found across Aotearoa

Peak season
March to August

Size
Can grow up to 4 cm tall

Calocera species could also be in the jelly fungi category. They're semi-gelatinous, but are erect and branching like other coral fungi. They grow from wet rotting logs and there are at least 10 species in Aotearoa. Occasionally you'll find hundreds all growing from the same log.

Goldclub coral

Scientific name
Clavulinopsis species

Distribution
Bay of Plenty, Waikato, Taranaki and Wellington in the NI, West Coast and Otago in the SI

Peak season
March to July

Size
Each branch can stand up to 5 cm tall and 3 mm wide

These yellow-orange spindles are fun to spot. Look for them in cooler months, growing in small groups among leaf litter and from rotten wood in pine and mānuka forests.

Wrinkled club fungus

Scientific name
Clavulina rugosa

Distribution
Gisborne, Taranaki, Manawatū, Wellington. Nelson, Tasman, Marlborough, Canterbury and Otago

Peak season
March to August

Size
Each branch can stand up to 8 cm high and 3 mm wide

This species is stark white, branching, and tends to grow in large groups. Look for it growing from pine needle beds. I've only ever seen this introduced species in urban areas where there are lots of exotic trees around.

Flame fungus

Scientific name
Clavulinopsis sulcata group

Distribution
Found across the NI and SI, apart from Canterbury and Otago

Peak season
March to August

Size
Each branch can stand up to 7 cm tall and 7 mm wide

Flame fungus has a bright pink-orange colour to it. It sometimes grows as a single flame, and sometimes as a full bonfire with several tendrils grouped together. Generally, when young, this fungus tapers towards the top, but over time the upper part of the tendril can become inflated and more club-shaped. Look for these tiny flames among leaf litter and at the base of tree ferns.

A *Beauveria* species that has parasitised a giant wētā.

Insect killers

Some fungi eat dead organic matter; others get their kai from plants. *This* group parasitises different types of insect. They use the body of the insect as food, killing it in the process. They are expert mummifiers, morticians, undertakers, embalmers — whichever name you want to use.

Mycorrhizal fungi remind us that survival of the fittest is not a 'one size fits all' way to describe the relationships that exist in nature — there are competitive *and* cooperative relationships. Parasitic fungi, on the other hand, remind us that those relationships are rarely balanced 50:50, and that sometimes one party stands to benefit far more than the other.

Some parasitic fungi are associated with an insect. And far more insects than I initially thought have a parasitic fungus lying in wait for them that sometimes causes their demise. Even giant wētā have a parasitic fungus associated with them. But wētā are vulnerable and dwindling in number, so their fungi spectres may be on the decline too.[28] There are also heaps of parasitic fungi associated with plants — they're mostly species with small fruiting bodies that can't be seen with the naked eye. Parasitic fungi even affect us. Ailments like athlete's foot are caused by fungi.

Āwhato

Scientific name
Ophiocordyceps robertsii

Other common names
Āwheto, hāwhato, hōtete, mokoroa, vegetable caterpillar

Distribution
Auckland, Waikato, Manawatū, Hawke's Bay and Wellington regions of the NI. Nelson, Tasman, Marlborough, West Coast and Otago regions of the SI

Peak season
October to March

Size
Stands about 10 cm tall and about 3 mm wide

Āwhato, a species endemic to Aotearoa, was the first fungus to be scientifically described here, back in 1836. It has been used as medicine, and even as ink for tā moko (tattooing). Look for it growing from the leaf litter, especially in the central North Island and upper South Island.

This species sneaks up on unlucky caterpillars as they munch away, gorging themselves on forest detritus in preparation to turn into moths. As the caterpillars go about their lives eating and burrowing, they occasionally encounter an āwhato spore. The spore gets trapped in the folds of the caterpillar's neck, or its mouth parts, and that's when the mummification begins.

From there, the fungus feeds on the caterpillar's body, and begins to fill it with mycelium. The fungus somehow manages to preserve the caterpillar's body so well that you can still make out the hairs, mouth parts and other fine details on the mummified caterpillar.

At the end of its lifespan, the fungus sends up a fruiting body from the caterpillar corpse. Straight out and up from its body. The fruiting body resembles a bulrush. It was reportedly sometimes dried, burned into a coal, mixed with animal fat and made into a black pigment for tā moko, and may have been preferred by some tohunga for its depth of pigment. There are also some reports that it was powdered and used as a medicine.[29] This strange-looking fungus was even sold as a souvenir in the past. Travellers were apparently happy to pay good money for the mummified bug and fungal growth.[30]

While parasitic *Cordyceps* species like this one are bad news for many insects, several species have been shown to have positive applications for humans. They are said to boost energy, increase lung capacity, and even increase blood oxygenation. In traditional Chinese medicine they are used to support virility, and many retailers now sell powdered *Cordyceps* capsules to consumers who want a natural brain boost.

Vegetable cicada

Scientific name
Cordyceps sinclairii

Distribution
Found across the NI apart from Hawke's Bay. In the SI it has been seen in Nelson, Tasman, Marlborough, West Coast, around Christchurch and in Southland

Peak season
December to June

Size
Stands about 4 cm tall and about 4 mm wide

This parasitic fungus can be found through summer and early autumn; it sprouts up from leaf litter. I've only seen it once, on the West Coast of the South Island. I didn't know what it was at the time. It looked like a miniature feather-duster with a slender pale stalk and light, powdery petal-like structures at the top. As I brushed the tops with my finger the spores made a *poof* of white haze. Unknowingly, I'd pre-empted the breeze and released an onslaught of parasitic spores on to the area's cicada community.

Cicadas spend the first phase of their life as underground nymphs. For three years they forage for tasty morsels in the soil. Every so often, as they're munching away, they hoover up a *Cordyceps sinclairii* spore. From there the spore takes hold. After lying in wait, its moment has arrived — it's ready to launch into the next phase of its own life. From that moment, the spore and the cicada's lives become terribly tangled.

As the cicada nymph continues to eat, the spore uses the nutrients to begin building a mycelial network within the nymph's body. Eventually, the fungi out-muscles the nymph, killing it. What remains is a mummified version of the nymph, buried in the soil. When conditions are right, the mycelial mummy will send up a fruiting body to spread more spores and ensnare more victims.

Despite its semi-wicked nature, medicine that helps to treat multiple sclerosis has been developed from a chemical compound that fungi in this genus contain. In addition to the species shown here, we also have *Cordyceps tenuipes* in Aotearoa, which looks near-identical but with yellow stems.

Icing sugar fungus

Scientific name
Beauveria species

Distribution
Northland, Auckland, Waikato, Bay of Plenty, Hawke's Bay, Taranaki, Manawatū and Wellington regions of the NI. Nelson, Tasman, West Coast, around Dunedin and in Southland in the SI

Peak season
January to June

Size
Varies depending on which insect it parasitises

This fungus reminds me of a well-practised baker. It manages to perfectly trace the delicate edges of bug exoskeletons with a bright white, crystalline fungi 'icing'. The result is a spooky birthday cake — or death cake, rather, since the process kills the bug.

Look for this fungus in summer and early autumn. You may spot the bug first, though. It's not uncommon to find icing sugar fungus where the victim remains somewhat visible and intact. These specimens look like little works of art: the bright white fungus is striking against the shiny, green and deep black armour of beetles and other bugs.

This fungus has been used as a natural insecticide to target everything from termites to whiteflies — those pesky bugs that eat up cruciferous vegetables in your garden like broccoli and kale.

Lichen

Aotearoa is home to a mind-boggling number of lichens. Go look at the side of your house, the pavement, a street sign, the teeny tiny crevices on the outside of your car — I can almost guarantee you'll find a lichen in one, if not all of these locations. Lichen are ubiquitous. They cover up to 7% of the Earth's surface.[31] And up to 10% of the global biodiversity of lichen may be present here in Aotearoa.

Lichens are an algae and a fungus (and now research suggests possibly also bacteria) living together in a mutualistic kind of way. They're what first got me into fungi; they are my oldest guides when it comes to learning about not only *what* lives in the ngāhere, but also *how* those organisms live.

Most lichens don't have common names, but there are a few well-known, recognisable ones out there. And even some funny ones. In 2019, Dr Allison Knight, a lichenologist at the University of Otago, uploaded a few lichen finds to the iNaturalist website with the common name 'sexy pavement lichen'. This species was apparently being sold overseas in capsules to treat erectile dysfunction. The name stuck, and made headlines, but this species is not recommended for consumption. Later that year, Knight and some of her colleagues identified a lichen that hadn't been described yet and named it *Ocellularia jacinda-arderniae*.

Lichens are beautiful, largely unknown, and full of important wisdom. They challenge our idea of the individual — is a lichen one organism, or is it multiple. And am I one organism? I feel more like a multitude when I consider the fungi, bacteria and other microbes that live on and in me.

Lichens also challenge what it means to be alive — they can slow their metabolic processes to almost zero in times without rain, then spring back when the first, welcome drops return. Lichens also teach us about life; the word 'symbiosis' wasn't commonly used until scientists needed a new way to describe two partner organisms living in collaboration.

Lichens so often capture people's fascination and flip what we 'know' to be true upside down, but despite the breadth of lichen's teachings, there are only a handful of lichenologists in Aotearoa. Who knows what these wonders will reveal to us next?

Now, don't fret if you come across a lichen that you can't name. There are three main types to categorise your finds with, listed below.

- **Crustose lichen** — Crustose lichens, the 'paint splash' types, don't have much of a 3-D quality and are almost one with whatever surface they grow on.
- **Foliose lichen** — Foliose lichens are slightly 3-D and have a leafy quality to them.
- **Fruticose lichen** — These lichens are very 3-D and have structural bits that stand up or branch out from the point at which the lichen connects to whatever it is growing from.

Angiangi

Scientific name
Usnea species

Other common names
Kohukohu, hawa, old man's beard (not to be confused with the noxious weed)

Distribution
Found across Aotearoa

Peak season
Always about

Size
Varies widely

Angiangi is an *Usnea* lichen used in rongoā Māori (traditional Māori medicine) for its powerful antibacterial properties. It is found throughout the ngāhere across the country year-round. Angiangi (this also means 'a gentle breeze') clings to branches and trunks, hanging from them in delicate, light-green wisps. The powerful compounds in this lichen can also be used to yield a bright yellow dye.[32] And the usnic acid it contains also offers valuable medicinal properties. Early Māori used it to cover wounds and treat itches, abrasions and other skin conditions, and it may have been combined with kōhia oil to make ointment.[33] Today, angiangi tinctures can be purchased from rongoā Māori practitioners.

Foliose lichen.

Crustose lich

Fruticose lichen.

Slime mould

Slime moulds aren't fungi. Which means that they also aren't moulds. The only part of the name that rings true is the slimy bit. I think slime moulds need a rebrand, but for now this is the only common name we've got for these simple-looking yet wildly sophisticated organisms.

Slime moulds have shocked, inspired and terrified humans for centuries and they deserve more than a passing mention in this book for several reasons. Once you start looking for fungi in earnest, you'll soon stumble across some slime moulds — they like similar environments. And while they aren't fungi, slime moulds can help us understand and almost visualise mycelium.

Slime moulds are sort of like amoebas. They're microscopic, single-celled, sac-shaped organisms. They spend most of their lives rolling around the earthly plane as lone rangers — tiny, squishy tumbleweeds with a singular mission: to find food. These tiny guys mostly feast on bacteria, yeasts and fungi in the soil. They come in every colour of the rainbow except true green, as they don't contain chlorophyll like plants do.

When food is scarce, and several slime mould cells cross paths, something peculiar happens. The singular cells fuse into one 'blob'. Each individual softens its cell membrane to join the larger group. Suddenly a multitude of individuals, all with their own genetics, become part of one singular 'body'. There's no limit to how many can join this party, which is called a plasmodium.

From there, the plasmodium begins to spread in all directions, looking for a snack. When it finds something yum, it takes note. When it finds something it doesn't like, such as bright sunlight or salt, it also takes note. As it searches its surroundings, the slime mould lays down chemical markers that communicate these learnings and helps the group home in on food. As a result, slime mould rarely retraces its path. At the end of its life, the plasmodium changes form and function one last time, and becomes a sporangium. It goes from slimy to crusty, and forms stalky fruiting bodies which produce dusty spores that are caught and spread by the wind. When those spores come to land, they'll try to establish themselves as singular slimy tumbleweeds, starting the journey once more.

While slime moulds don't have a central nervous system or anything even a little bit like one, they can efficiently solve very complex problems. The study that started to generate some modern-day hype around them is known as the 'Tokyo study'.

In 2019, I went to Tokyo by myself. Navigating the city's subway system was really tricky. After about 10 days of ballsing it up, I learned it was much faster to ask someone for directions because, while it's complex, the Tokyo system is very clever. Those who know how to use it can quickly navigate this city of 14 million.

In the Tokyo study, researchers at Hokkaido University placed one of the most well-known slime mould species into a scaled-down 'maze' replicating the Tokyo train system. They placed tasty oat snacks at major 'hubs', and then . . . waited. In just 24 hours, the slime 'determined' the fastest way to get between each of the snack stations.[34] The route that the slime mould 'chose' was the same route humans in the know would take.

This study validated two things: first, the Tokyo system is highly efficient if you use it right; and second, slime moulds can build highly efficient connective networks. Similar studies have been conducted with transport networks from Canada, the United States and the United Kingdom, as well as the Silk Road trade route. Slime mould navigated all of these with peak efficiency, re-creating the same routes that our human brains required equations, algorithms, and trial and error to produce.

Slime moulds are not mycelium, but they do share some intriguing traits. Mycelium lives underground, within wood or other substrates. Slime moulds like similar environments, and both are keen foragers. They're clever too, both reinforce network links that lead to food and prune back ones that don't. Both are 'feeding systems' that solve the arduous yet universal need that non-plants have, to find food.

Both slime moulds and mycelium are wildly thought-provoking. They make a strong case for brainless organisms being able to interpret their environments, respond, communicate and problem-solve at scale. Like lichen, they also challenge our idea of the individual: is a slimy plasmodium one individual or many? Likewise, is mycelium, with its constant movement and expansion, an organism or a process?

Scientists are even using slime moulds to create new system networks. Because they are small, portable and easy to work with, slime moulds can be used to model things like road repair closures before they happen. In 2020, astronomers used slime mould to 'map' dark matter — the cosmic dark web of filaments that links distant galaxies together. I cannot even begin to wrap my head around that, but slime moulds, it would seem, absolutely can.[35]

A white slime mould.

Dog vomit

Scientific name
Fuligo septica

Other common name
Scrambled egg slime

Distribution
Found across Aotearoa

Peak season
October to April

Size
Usually 25–200 mm in diameter and 10–30 mm thick

This bright yellow slime is the most common one, in my experience. It loves to grow in mulched garden beds, alongside park paths, and in other areas with high human traffic. It can also grow quite huge. I've seen specimens larger than both my hands put together. There's no need to look too hard for this one — it grows up and down the country, and will probably find you first.

In folklore from various parts of the world, this slime species has been explained as troll cat throw-up, witches' butter and witches' spit. It can rapidly colonise mulch beds, but in the bush I've seen it grow on the leaves and stems of living plants, as well as across mossy expanses. When it enters the sporangium stage it changes colour and texture, becoming buff-coloured and dusty.

Chocolate tube

Scientific name
Stemonitis splendens

Distribution
Found across Aotearoa

Peak season
November to March

Size
Each sporangium is 10–20 mm tall and 1–2 mm across, with a stem 3–5 mm long and less than 1 mm thick

This pretty common, pretty cool slime mould species features upright tubes that are clustered together. They almost look like chocolate ice blocks. They have slender dark stems and powdery brown spore surfaces. Look for this slimy species on sheltered dead wood. They can be tough to spot against brown woody debris, so look closely, and get out your best forager squat to check the sides of dead logs, too.

Wolf's milk

Scientific name
Lycogala epidendrum

Distribution
Grows across Aotearoa apart from Marlborough, South Canterbury and Central Otago

Peak season
October to May

Size
Each blob is about 3–15 mm across

Look for this bright-pink slime on dead, damp, rotting wood. The pink puffs, which are sometimes orangey, turn solid grey as they enter the sporangium stage and it's not uncommon to see both life phases on the same log. Press one of the blobs and you'll find it pops like a pimple — yuck!

Salmon eggs

Scientific name
Trichia decipiens

Distribution
Auckland, Bay of Plenty, Waikato, Hawke's Bay and Wellington in the NI. Nelson, Tasman, Canterbury, Southland and Otago in the SI

Peak season
April to October

Size
Each egg is up to 3 mm high and 0.6–0.8 mm wide

This pinky-orange slime mould is found on dead conifer wood. Each 'egg' sits atop a small stem that is lighter in colour. You can often find it in masses of many hundreds clustered together, and it is also sometimes called beaded slime mould.

Honeycomb coral

Scientific name
Ceratiomyxa fruticulose

Distribution
Found throughout Aotearoa

Peak season
Always around

Size
Individual fruiting bodies are 0.5–1 mm wide and 1–10 mm high

A common slime mould in Aotearoa. It has a bright-white, almost frosted colour and a coral-like shape. Look for it on water-logged dead wood in most forest types.

Pretzel slime mould

Scientific name
Hemitrichia serpula

Distribution
Found mostly in the NI, in Northland, Auckland, Bay of Plenty, Taranaki and Hawke's Bay

Peak season
August to December

Size
Each strand is just 0.4–0.6 mm wide

This slime mould has a unique geometric aesthetic. It has a bright orange colour and distinctive pattern. Very pleasing to the eye. When it gets to this hardened, more mature stage, it takes on a shiny, hard texture like a pretzel. Look for it on rotten logs.

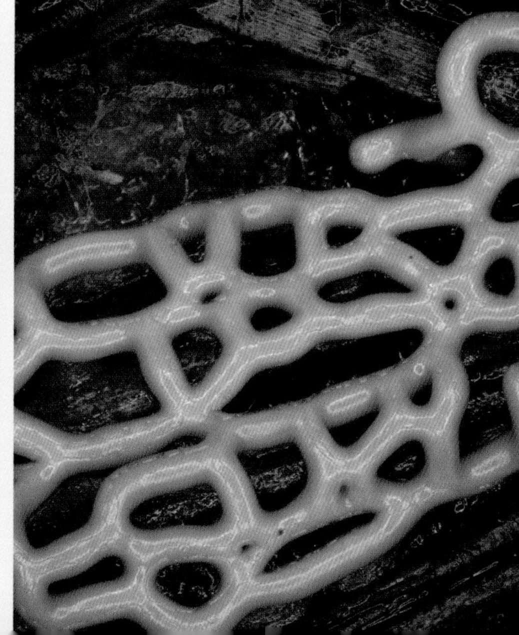

A final word on fungi

When I started writing this book, I thought I'd learn heaps about fungi. And I did. But I also learned a lot about wonder.

There's a whole heap of wonder in this world. And whoever created it packed a tonne into Aotearoa. From fiords to ferns, from wētā to werewere-kōkako, awe is around every corner here. When the noise of life manages to drown out that magic, fungi-spotting is a fantastic way to find it again.

Fungi inject wonder into my world, often when I least expect it. How magical is it when you think of fungi and then spot one on the way to work? How joyous is it when you finally find an edible mushroom? How absolutely amazing is it that after seeing one fungus, a thousand more pop out at you? How brain-bending is it to consider their mycelial minds? I'd love to see a brain scan of myself before and after spotting a colourful fungus. I imagine those scans would light up like a bioluminescent mushroom.

The practice of fungi-spotting often feels like a game without end or time outs. Hopefully you're now playing along, noticing fungi, possibly even foraging them, cooking with them, or growing some in the garden. To play this wonderful game, all you have to do is get going. Just a five-minute walk down your street can become a fungi-spotting affair. Pointing a fungus out to a friend is all it takes to bring them into the fold.

These micro-adventures have become a time to reconnect with my senses. To notice the flora, fauna and fungi in even the most mundane spaces. To be among the birdsong and feel the sun on my skin. To get off my screens, out of my head and back into the present moment.

Fungi peel back a layer of the day-to-day. Spotting one

creates a little pocket of space–time where something surprising can happen. In that moment of noticing, the real world falls away. The ritual of looking for fungi is powerful. It's a zero-cost tool that I will always have with me.

Within that little pocket of space-time, in that moment of magic, there's a lot of mystery too. The story of fungi is full of gaps. We know so little about this vast kingdom. What knowledge of fungi has been lost? What knowledge is yet to emerge? While I was writing this book, a new albeit speculative study came out that suggested fungi can 'talk'. The study measured electrical signals moving through a mycelium. This mycelium used 50 different spikes in electrical activity to communicate across the network. These spikes, the researchers suggested, resemble 'words' and their many combinations 'sentences'. Like below-ground gossipers, fungi are chattering away beneath our feet. They invite us to learn their language, and we might do well to learn it fast.

Fungi are resilient as heck, but at the current rate we're going it'll take ages to describe them all. In that time, many could become extinct before we even know they're there. If we can buy ourselves some time, look after them a bit better, who knows what solutions, foods, medicines and other magic they might share with us?

There's so much you can do for fungi. Share them with a friend. Take your mate on a foray. Log what you find on iNaturalist. The more people interested in fungi, the better. And remember: many of the most important things we have learned from fungi have been uncovered simply by paying attention to them. Scientist or regular person, fungi-fanatic or fresh to the game — it doesn't matter. You can make an impact. You've now got a new lens. Fungi glasses to peer out at the world through.

I can't wait to see what you find with these glasses. New fungi-based solutions are surely yet to be revealed. And, equally as important, new perspectives. Fungi change how we look at the world and have the power to change our minds. They remind us that we're all connected, and part of a big, beautiful, tangled web of life. They have the power to wow us, inspire us, heal us. Paying attention to fungi is sometimes nothing short of radical. It can change you and change the world. In fact, it already has.

Thanks

Many thanks and mush love to everyone who has supported this book. I am deeply grateful to:

To Marty Jones, for gifting me the grain of sand that this book was built around.

To my friends at Foundation Café, especially Suz and Anne. For your friendship, flat whites and encouraging my 'mushroom woman' persona.

To my family. Mom, Dad, Phoebe and Denison for being the best spotters of small wonders. For teaching me how to look and for looking with me . . . for sea glass, starfish, fungi, frogs. I adore this pastime we share. Thank you for your love.

To Julia Keogh-Cope, Poppia Marriott and all the good folk at Tūranga Central Library who helped me forage for stories and info as I wrote this. Tūranga has been a bright spot in my life since the day it opened in 2018. Much of this book was written there.

To my work whānau at New Zealand Trade & Enterprise, thank you for encouraging me.

To Shirley Kerr, Peter Buchanan, Jerry Cooper and Peter Langlands — thank you for bringing your expertise to this book and for making it better than I could have on my own. Thank you for the work you do in the world of fungi.

To the Wood family, thank you for celebrating me and looking after me.

To my friends who make up my mycelial network, which expands through space and time in a way I never could've imagined. From America to Aotearoa, some of you have been with me for more than 20 years. Thank you Meghan, Lindsay, Hayley, Allie, Leah, Kendall, the Great Eight, Wes, Liz, James, Will, Lauren, Jess, Amon, Angus, Oscar, Molly, Alba, Tess, Anthony, Beth, the Exiles, the Paroa Gang and Whitby Street. I adore you all.

To the Eat New Zealand kaitaki. Angela Clifford, Kate

Underwood, Aliesha McGilligan, Max Gordy, Vicki Young, Lucas Parkinson and all my mates within that organisation. Thank you for sharing your food (and fungi) froth with me.

To Sophie Merkens, a guiding light on this book-writing rollercoaster. Thank you for helping me shape this idea, and for five-minute voice messages in uncertain moments.

To my teachers, especially Leigh-Ann Beavers, Lisa Greer, Jennifer Bonafide and Ryan Jackson. I hope you can hear your own voices in these pages.

To the many people who shared their fungi wisdom, research, businesses and memories with me. Joe McLeod, Genevieve Early, Chris Smith, James Ferrier, David Hera, Mitchell Head, Rachael Sumner, Gareth Renowden, Jim Bob Fuller, Jonathan Thevenard, Silas Villas-Boas, Taylor McConnell, Susan Jacobs, Kelly Styles, Emily Blanchett and Fionnuala Bulman.

To Paula Vigus, your fantastic photos made this book come alive. And made it more beautiful than I thought possible. You're a legend.

To Nancy Zhou, for encouraging my writing to this stage. Very few people have been to as many regional NZ cafés as we have. Thanks for all the fun times, and for capturing my foraging joy so perfectly for this book.

To Rachel Eadie, for bringing your immense talent to this and for lifting me up when a fun book about fungi felt hard. For helping me bring a childhood dream to reality. Thank you for cheering me on and pushing me hard all at once.

To the many Penguins who worked on the book, including Stuart Lipshaw, Carla Sy and Claire Murdoch, and also my editor Teresa McIntyre — thank you for filling our lives with beautiful books crafted with care.

To Duncan, my partner. For nourishing meals, gentle pats and kind words along the way. You bring so much fun and softness to my life. I could not have done this without you. Thank you.

Image credits

Paula Vigus: 2, 4–5, 6, 14–15, 16, 19, 39, 44, 51, 52–53, 54, 61 (bottom), 63 (top), 64, 65 (top), 66 (bottom), 67, 75, 77, 78–79, 101, 115 (top), 150–51, 156, 159, 163 (top), 165, 167, 168, 170, 172, 174, 177, 178 (right), 179, 180, 181, 183, 184, 185, 186, 187, 189, 190 (top), 198, 200 (left), 201 (left), 203, 206, 208, 211, 213 (bottom), 215 (bottom), 217, 218 (bottom), 220, 223, 225 (right), 226, 227, 233, 234 (top), 239 (bottom), 241, 242, 244, 246 (right), 247 (left), 248, 251, 253, 255, 257, 263 (right), 264, 273, 275, 277, 278, 280 (right), 281 (right), 282, 285 (top), 289, 290 (top), 292, 296, 303, 313, 315 (bottom), 336

Nancy Zhou: 8, 59, 87, 93, 94–95, 318, 321, 335

Liv Sisson: 13 (illustrations), 21, 25, 27, 31, 61 (top), 62, 63 (bottom), 65 (bottom), 66 (top), 80, 85 (top middle), 96, 102–103, 109, 113, 114, 115 (bottom), 117, 118, 119, 121, 123, 125, 127, 128, 129, 131, 133, 137, 143, 144, 145, 152, 155, 163 (bottom), 229, 239 (top), 246 (left), 258 (bottom), 262 (left), 267, 280 (left), 287 (left), 301, 306, 308, 309, 315 (top), 316 (left)

iStock: 36, 85 (middle left and right, bottom left), 111, 147, 149, 161, 192, 201 (right), 204, 215 (top), 234 (bottom), 236, 262 (right), 269, 295 (left), 316 (right), 317

Shirley Kerr: 85 (top left, top right), 105, 107, 190 (bottom), 195, 200 (right), 213 (top), 224, 225 (left), 231, 247 (right), 260, 263 (left), 281 (left), 285 (bottom), 286 (right), 287 (right), 290 (bottom), 294, 299

Shutterstock: 85 (bottom right), 135, 139, 141, 178 (left), 197, 218 (top), 295 (right)

Alamy: 258 (top, Sheryl Watson), 271 (Robert Wyatt), 286 (left, Jelger Herder)

Additional resources

These resources are welcome additions to any fungi fan's bookshelf/digital library. My copies of these titles are well-worn and many of them have been used and referenced throughout this book. Many thanks to the authors — you have added so much to my understanding of this wondrous world.

A Field Guide to New Zealand Fungi
An extensive and colourful field guide by Shirley Kerr, who has contributed heaps to our understanding of what fungi can be found in Aotearoa. I love this guide and use it often.

Braiding Sweetgrass* and *Gathering Moss
Two books by Robin Wall Kimmerer, which look at the teachings of the natural world and the power of looking through the lens of both indigenous wisdom and scientific knowledge.

Entangled Life
A book by Merlin Sheldrake that dives into how fungi 'make our worlds, change our minds and shape our futures'. Utterly fascinating.

Finding the Mother Tree
A book by Suzanne Simard on 'the wisdom of the forest' and the fungal communication networks they use.

How To Change Your Mind
A book by Michael Pollan on 'what the new science of psychedelics teaches us about consciousness, dying, addiction, depression and transcendence'. Again, utterly fascinating.

iNaturalist.nz
A website where you can record what you see in New Zealand nature and learn more about what's here.

Mushrooms and Other Fungi of New Zealand
A pocket field guide by Geoff Ridley and Don Horne. An excellent first look at some of the unique species we have here as well as where and when to find them.

New Zealand's Virtual Mycota
A site hosted by Manaaki Whenua Landcare Research, which is a useful identification tool. Find it at virtualmycota.landcareresearch.co.nz.

Ngā Hekaheka o Aotearoa
A teaching guide created by Peter Buchanan, Georgina Stewart and Hēni Jacob, which details the forms, functions and importance of fungi found in Aotearoa, including examples of Māori knowledge and use of fungi. Download this for free at huia.co.nz.

Te Mahi Oneone Hua Parakore
A Māori soil sovereignty and wellbeing handbook by Dr Jessica Hutchings, which highlights Māori relationships with soil and the connections between our soil, our food system and our health.

The Forager's Treasury
A book by Johanna Knox, which is my most favourite foraging guide. While it doesn't include fungi, it does include loads of species that you can spot and learn about while in pursuit of edible fungi.

The Hidden Forest
A website curated by Clive Shirley that features a huge range of fungi species as well as photos and field-guide information.

Wild Capture
A Facebook page hosted by expert forager, Peter Langlands. An excellent source for NZ specific, seasonal foraging information. Peter also has a range of foraging guides, which can be purchased on TradeMe.

Glossary

Algae (singular = alga): an informal term for a large and varied group of photosynthetic organisms that are eukaryotic, meaning their cells have nuclei. Many algae are single-celled, but macro algae like seaweed also exist. Different to plants, algae do not have stems, leaves or true roots.

Cap: the top of a mushroom. Also called a pileus.

Coprolite: fossilised poo.

Cortina: the cobweb-like partial veil on some mushrooms, particularly mushrooms in genus *Cortinarius*. The fibres of the cortina sometimes hang from the edge of the cap or the stipe after the mushroom has opened.

Ectomycorrhizal: describes fungi that do not penetrate their host plant's cell walls. Instead these fungi form intercellular points of contact, called Hartig nets, which are made up of branching hyphae that form a net structure between root cells.

Endemic species: species that are restricted to a certain area of the globe and do not occur anywhere else.

Endophyte/endophytic fungi: grow within plant tissue, penetrating the plant's cells and helping it absorb nutrients and resist disease.

Fibrils: a hairy texture on the caps of some mushrooms.

Gasteroid: secotioid fungi that have evolved to no longer have stipes. Only the pouch structures, which hold spores internally, remain.

Gills: a thin, ribbed texture found underneath the caps of some mushrooms. Also called lamella.

Hyphae: the branching threads that together make up mycelium. Hyphae wander through the soil of whatever growing substrate they are in, searching for nutrients. This is one way a fungus grows.

Mushroom spawn: a substrate that already has mycelium growing on it. Used to inoculate or 'add spawn' to garden beds where you might want to grow mushrooms.

Mycelium: the root-like part of a fungus that occasionally produces a reproductive fruiting body like a mushroom. Mycelium is made up of fine, branching white threads called hyphae. Mycelium is found in and on soil, on dead wood and other substrates.

Mycoheterotrophs: plants that are parasites of fungi.

Mycologist: a scientist who studies fungi.

Mycorrhizal: describes a symbiotic association between a fungus and a plant. In a mycorrhizal relationship the fungus colonises the host plant's roots.

Native species: a species that is present in a certain geographical location only as the result of local natural evolution.

Periodioles: the tiny 'eggs' found within bird's nest fungi that contain the reproductive spores.

Plasmodium: a form within a slime mould's life cycle where many singular slime mould cells group together into a larger slimy mass.

Podocarps: unique conifer trees. Like all conifers, podocarps use cones to reproduce, but podocarp cones look more like berries. These attract birds who then help spread the seeds. The most well-known podocarps in Aotearoa are 'the giants': miro, rimu, kahikatea, mataī and tōtara.

Broadleaf podocarp forests: this term refers to mixed broadleaf podocarp forests, which host many cool fungi. The podocarps include but are not limited to miro, rimu, kahikatea, mataī and/or tōtara trees. The shorter broadleaf trees that tend to grow below the podocarps include but are not limited to puahou five-finger, karamū, horoeka lancewood, rautāwhiri, juvenile rewarewa honeysuckle, juvenile kāmahi, māhoe, kōtukutuku tree fuchsia, kōtētē seven finger as well as mamaku and ponga tree ferns.

Polypores: a group of fungi that have large fruiting bodies and pores (lots of tiny openings) on their undersides.

Reticulation: a lacy pattern seen on the stipes of some mushrooms, like porcini.

Saprobe/saprophytic fungi: fungi that feed on dead wood, animal remains and other organic matter. These fungi break down dead stuff and turn it into rich soil and usable nutrients for plants.

Secotioid: gilled mushrooms that have evolved to have gill-less pouch-shaped caps.

Skirt/ring: the remnants of a partial veil that hang from some mushrooms' stipes. Skirts are generally more prominent whereas rings are smaller.

Sporangium: a form within a slime mould's life cycle where a plasmodium transforms into rigid fruiting bodies that release reproductive spores.

Spore: a very small, usually singularly celled reproductive unit that can generate a new individual of a fungus.

Spore surface: the part of a fungi that holds the reproductive spores. For mushrooms, this is the gills or pores. Also called a hymenophore.

Squamules: small scales found on the caps of some mushrooms.

Stain: some fungi change colour when bruised, this is called a stain.

Stipe: the stem of a fungi, usually refers to a mushroom stem.

Striations: long, thin parallel lines, streaks or stripes.

Substrate: soil, dead wood, leaf litter, organic matter — a surface or material through which fungi grows and obtains nutrients.

Symbiont: an organism living in symbiosis with another.

Symbiosis: any type of close and long-term relationship between two organisms.

Umbo: a bump in the centre of some mushroom caps, which can be a helpful identification feature in some cases.

Universal veil: a temporary tissue membrane that completely encases some mushrooms when they first emerge. As the mushroom opens it breaks the universal veil.

Partial veil: a temporary tissue membrane that initially covers the gills or pores of some mushrooms and protects the spores held there as they mature. As the mushroom matures and expands, the partial veil tears and falls away. It sometimes then hangs from the stipe and is called a skirt.

Vascular: describes plant tissues that conduct water and nutrients within flowering plants and ferns.

Volva/egg sac: a cup-like structure at the base of some mushrooms that is the remnant of a universal veil.

Endnotes

What are fungi?

1 Anne Casselman, Strange but true: The largest organism on Earth is a fungus. *Scientific American*, 4 October 2007.

2 David L. Hawksworth and Robert Lücking, Fungal diversity revisited: 2.2 to 3.8 million species, *Microbiology Spectrum*, vol. 5, no. 4, 2017, doi: 10.1128/microbiolspec.FUNK-0052-2016.

3 Merlin Sheldrake, *Entangled Life: How fungi make our worlds, change our minds and shape our futures*, Random House, 2020, p. 232.

4 Ahmed Abdel-Azeem, Mohamed Abdel Azeem, Robert Blanchette, Marwa twakol Mohesien and Fatma Mahmoud Salem, The conservation of mushroom in Ancient Egypt through the present. Paper presented at The First International Conference on Fungal Conservation in the Middle East and North of Africa, Ismailia, Egypt, 18–20 October 2016.

5 Hannah Vickers, What is a fairy ring and what causes them? Woodland Trust blog, 26 August 2019.

6 Aparna Vidyasagar, Facts about the fungus among us, *Live Science*, 5 February 2016.

7 S. L. Baldauf and J. D. Palmer, Animals and fungi are each other's closest relatives: Congruent evidence from multiple proteins, *Proceedings of the National Academy of Sciences of the United States of America*, vol. 90, no. 24, 1993, pp. 11558–62.

8 Serita D. Frey, Mycorrhizal fungi as mediators of soil organic matter dynamics, *Annual Review of Ecology, Evolution, and Systematics*, vol. 50, 2019, pp. 237–259.

9 Toby Kiers and Merlin Sheldrake, A powerful and underappreciated ally in the climate crisis? Fungi, *The Guardian*, 30 November 2021.

10 Toby Kiers and Merlin Sheldrake, Key messages, Global Symposium on Soil Erosion (GSER19), Rome, Italy, 15–17 May 2019, Food and Agriculture Organization of the United Nations, https://www.fao.org/about/meetings/soil-erosion-symposium/key-messages/en/

11 Nina Lakhani, Alvin Chang, Rita Liu and Andrew Witherspoon, Our food system isn't ready for the climate crisis, *The Guardian*, 14 April 2022.

12 Peter Buchanan, Georgina Stewart and Hēni Jacob, *Ngā Hekaheka o Aotearoa — He aratohu mā te Pouako* (*The Fungi of Aotearoa — A guide for teachers*), Manaaki Whenua, 2017, p. 30.

13 Mary Jo Feeney, Amy Myrdal Miller and Peter Roupas, Mushrooms — biologically distinct and nutritionally unique: Exploring a 'third food kingdom', *Nutrition Today*, vol. 49, no. 6, 2014, pp. 301–307.

14 A. Mitchell and G. P. Savage, *Agrocybe parasitica*: The mushroom of the future?, *Proceedings of the Nutrition Society of New Zealand*, vol. 15, 1990, pp. 175–178.

15 Jessica Hutchings and Jo Smith (editors), *Te Mahi Oneone Hua Parakore: A Māori Soil Sovereignty and Wellbeing Handbook*, Harvest: Fresh Scholarship from the Field, 2020, p. 54.

16 *Ibid.*

17 Min Jia, Ling Chen, Hai-Liang Xin, Cheng-Jian Zheng, Khalid Rahman, Ting Han and Lu-Ping Qin, A friendly relationship between endophytic fungi and medicinal plants: A systematic review, *Frontiers in Microbiology*, 9 June 2016.

18 J. R. Leake and D. J Read, Mycorrhizal symbioses and pedogenesis throughout Earth's history, in Nancy Collins Johnson, Catherine Gehring and Jan Jansa (eds), *Mycorrhizal Mediation of Soil*, Chapter 2, pp. 9–33, Elsevier, 2017.

19 Mark C. Brundrett and Leho Tedersoo, Evolutionary history of mycorrhizal symbioses and global host plant diversity, *New Phytologist*, vol. 220, no. 4, 2018, pp. 1108–1115.

20 Carbon sequestration in soils, Ecological Society of America, 2000.

21 Serita D. Frey, Mycorrhizal fungi as mediators of soil organic matter dynamics, *Annual Review of Ecology, Evolution, and Systematics*, vol. 50, 2019, pp. 237–259.

22 Forests and agriculture, European Commission, Climate Action, https://climate.ec.europa.eu/eu-action/forests-and-agriculture_en

23 Merlin Sheldrake, *Entangled Life*, p. 168.

24 Yuan Song, Suzanne W. Simard, Allan Carroll, William W. Mohn and Ren Sen Zeng, Defoliation of interior Douglas-fir elicits carbon transfer and stress signalling to ponderosa pine neighbors through ectomycorrhizal networks, *Scientific Reports*, vol. 5, 2015, art. no. 8495.

25 Dave Hansford, The Wood Wide Web, *New Zealand Geographic*, issue 148, Nov–Dec 2017.

26 Dave Davies, Trees talk to each other. 'Mother tree' ecologist hears lessons for people, too, *Shots: Health News from NPR*, 4 May, 2021.

27 Merlin Sheldrake, *Entangled Life*, p. 153.

28 Facts on plastic waste, ecotricity, 20 January 2019, https://ecotricity.co.nz/facts-on-plastic-waste

29 Wilding conifer control in NZ, Ministry for Primary Industries Manatū Ahu Matua, https://www.mpi.govt.nz/biosecurity/long-term-biosecurity-management-programmes/wilding-conifers/

30 J. M Thwaites, R. L. Farrell, S. D. Duncan, R. T. and R. B. White, Fungal-based remediation: Treatment of PCP contaminated soil in New Zealand, in S. N. Singh and R. D. Tripathi (eds), *Environmental Bioremediation Technologies*, 2007, Chapter 20, pp. 465–480, Springer.

31 Matthew I. Hutchings, Andrew W. Truman and Barrie Wilkinson, Antibiotics: Past, present and future, *Current Opinion in Microbiology*, vol. 51, 2019, pp. 72–80, https://www.sciencedirect.com/science/article/pii/S1369527419300190

32 Maryn McKenna, When we lose antibiotics, here's everything else we'll lose too, *Wired Science*, 28 November 2013, https://www.wired.com/2013/11/end-abx/

33 Henry T. Tribe, The discovery and development of cyclosporin, *Mycologist*, vol. 12, no. 1, 1998, pp. 20–22, https://www.sciencedirect.com/science/article/abs/pii/S0269915X98801006

34 Māori knowledge and use of fungi, Science Learning Hub Pokapū Akoranga Pūtaiao, https://www.sciencelearn.org.nz/resources/2668-maori-knowledge-and-use-of-fungi

35 Ko Aotearoa Tēnei, Te Taumata Tuatahi, A Report into Claims Concerning New Zealand Law and Policy Affecting Māori Culture and Identity, Wai 262, Waitangi Tribunal Report, 2011, p. 2.

36 Siouxsie Wiles, The world is desperate for new antibiotics, and New Zealand's unique fungi look promising, *ABC News*, 9 September 2021.

37 Robin Young and Samantha Raphelson, One mycologist on why fungi are 'critical for the survival of life on this planet', WBUR, 28 January 2019, https://www.wbur.org/hereandnow/2019/01/28/mushrooms-fungi-disease-bees

38 Dalia Akramiene, Anatolijus Kondrotas, Janina Didziapetriene and Egidijus Kevelaitis, Effects of beta-glucans on the immune system, *Medicina (Kaunas)*, vol. 43, no. 8, 2007, pp. 597–606.

39 Michael Lim and Yun Shu, *The Future is Fungi*, Thames Hudson, 2022, p. 92.

40 Michael Pollan, *How to Change Your Mind*, Penguin Press, 2018, p. 320.

How to find and identify fungi

1 Robin Wall Kimmerer, *Braiding Sweetgrass: Indigenous wisdom, scientific knowledge and the teachings of plants*, Milkweed Editions, 2015, p. 49.

2 Peter Buchanan, Georgina Stewart and Hēni Jacob, *Ngā Hekaheka o Aotearoa — He Aratohu mā te Pouako* (*The Fungi of Aotearoa — A guide for teachers*), Manaaki Whenua, 2017.

3 Te Aka Māori Dictionary, https://maoridictionary.co.nz

4 Peter Buchanan, Georgina Stewart and Hēni Jacob, *Ngā Hekaheka o Aotearoa*.

How to forage for fungi

1 Merlin Sheldrake, *Entangled Life: How fungi make our worlds, change our minds and shape our futures*, Random House, 2020, p. 60.

Edible Fungi

1 Alan Clarke, *The Great Sacred Forest of Tane*, Reed, 2007, p. 196.

2 Peter Buchanan, Georgina Stewart and Hēni Jacob, *Ngā Hekaheka o Aotearoa — He Aratohu mā te Pouako* (*The Fungi of Aotearoa — A guide for teachers*), Manaaki Whenua, 2017, p. 30.

3 *Ibid.*

4 Vikineswary Sabaratnam, Wong Kah-Hui, Murali Naidu and Pamela Rosie David, Neuronal health — Can culinary and medicinal mushrooms help?, *Journal of Traditional and Complementary Medicine*, vol. 3, no. 1, 2013, pp. 62–68.

5 Peter Buchanan, Georgina Stewart and Hēni Jacob, *Ngā Hekaheka o Aotearoa*, p. 28.

6 *Ibid.*, p. 32.

7 Stephen Brightwell, Feasting on Fungi, *New Zealand Geographic*, issue 18, April–June 1993.

8 https://www.wildfooduk.com/mushroom-guide/shaggy-parasol/

9 Shirley Kerr, *A Field Guide to New Zealand Fungi*, self-published, 2019, p. 179.

Fungi of Aotearoa

1 Alison Ballance, Truffle-like fungi: What their genes can tell us, *Our Changing World*, RNZ, first broadcast 10 March 2016.

2 International recognition for New Zealand's endangered fungi, Manaaki Whenua news, 6 May 2020.

3 Shirley Kerr, *A Field Guide to New Zealand Fungi*, self-published, 2019, p. 59.

4 Winner picked for New Zealand's national fungus, Manaaki Whenua press release, 8 June 2018, https://www.scoop.co.nz/stories/SC1806/S00019/winner-picked-for-new-zealands-national-fungus.htm

5 Shirley Kerr, *A Field Guide to New Zealand Fungi*, p. 29.

6 *Laccaria*, Manaaki Whenua, https://virtualmycota.landcareresearch.co.nz/webforms/vM_Species.aspx?pk=5077

7 Sujal S. Phadke, Sex begets sexes, *Nature Ecology & Evolution*, vol. 2, 2018, pp. 1063–1064.

8 K. M. J. de Mattos-Shipley, K. L. Ford, F. Alberti, A. M. Banks, A. M. Bailey and G. D. Foster, The good, the bad and the tasty: The many roles of mushrooms, *Studies in Mycology*, vol. 85, 2016, pp. 125–127.

9 Erika Kothe, Mating types and pheromone recognition in the homobasidiomycete *Schizophyllum commune*, *Fungal Genetics and Biology*, vol. 27, nos. 2–3, 1999, pp. 146–152.

10 Monika Mahajan, *Schizophyllum commune*, *Emerging Infectious Diseases*, vol. 28, no. 3, 2022, p. 725.

11 Shirley Kerr, *A Field Guide to New Zealand Fungi*, p. 4.

12 Mateja Lumpert and Samo Kreft, Catching flies with *Amanita muscaria*: Traditional recipes from Slovenia and their efficacy in the extraction of ibotenic acid, *Journal of Ethnopharmacology*, vol. 187, 2016, pp. 1–8.

13 The interesting connection between Christmas and mushrooms, fantasticfungi.com, 11 December 2020, updated 14 September 2022.

14 Anne Casselman, Strange but true: The largest organism on Earth is a fungus, *Scientific American*, 4 October 2007, https://www.scientificamerican.com/article/strange-but-true-largest-organism-is-fungus/

15 Peter Buchanan, Georgina Stewart and Hēni Jacob, *Ngā Hekaheka o Aotearoa — He Aratohu mā te Pouako* (*The Fungi of Aotearoa — A guide for teachers*), Manaaki Whenua, 2017, p. 14.

16 Alison Ballance, Truffle-like fungi: What their genes can tell us.

17 *Russula kermesina*, http://www.hiddenforest.co.nz/fungi/family/rusulaceae/rusul09.htm

18 *Chlorociboria*, Manaaki Whenua, https://virtualmycota.landcareresearch.co.nz/webforms/vM_Species.aspx?pk=1590

19 Jennifer Frazer, Darwin's neon golf balls, *Scientific American*, 15 January 2013.

20 Joe Pinkstone, 'Slightly sweet' fungus that 'tastes of mucus' first discovered by Charles Darwin is unearthed in a pickle jar hidden in a basement 180 years after his voyage aboard the HMS *Beagle*, *Daily Mail*, 13 February 2019.

21 Peter Buchanan, Georgina Stewart and Hēni Jacob, *Nga Hekaheka o Aotearoa*, p. 17.

22 Te Aka Māori Dictionary, https://maoridictionary.co.nz

23 Allana Akhtar, 5 'functional' mushrooms the wellness industry is obsessed with, from lion's mane to turkey tail, *Insider*, 8 April 2022.

24 Sissi Wachtel-Galor, John Yuen, John A. Buswell and Iris F. F. Benzie, *Ganoderma lucidum* (lingzhi or reishi), Chapter 9 in I. F. F. Benzie and S. Wachtel-Galor (editors), *Herbal Medicine: Biomolecular and clinical aspects*, 2nd edition, CRC Press/Taylor & Francis, 2011.

25 Ko Aotearoa Tēnei, Te Taumata Tuatahi, Wai 262, p. 2.

26 Murdoch Riley, *Māori Healing and Herbal*, Viking Sevenseas NZ, 1994, p. 383.

27 Peter Buchanan, Georgina Stewart and Hēni Jacob, *Nga Hekaheka o Aotearoa*, p. 13.

28 Critter of the week: Vegetable caterpillar, from Afternoons with Jesse Mulligan, Radio New Zealand, 21 July 2017.

29 Peter Buchanan, Georgina Stewart and Hēni Jacob, *Nga Hekaheka o Aotearoa*, p. 11.

30 Critter of the week: Vegetable caterpillar, Radio New Zealand.

31 Johan Asplund and David A. Wardle, How lichens impact on terrestrial community and ecosystem properties, *Biological Reviews*, vol. 92, no. 3, 2017, pp. 1720–1738.

32 Alan Clarke, *The Great Sacred Forest of Tane*, Reed, 2007, p. 211.

33 Murdoch Riley, *Māori healing and herbal*, p. 120.

34 Katherine Harmon, Slime mold validates efficiency of Tokyo rail network, *Scientific American*, 21 January 2010.

35 Patricia Fara, The surprisingly exciting story of the woman who studied slime moulds, *Prospect*, 21 August 2021.

Index

Page numbers in italics indicate photographs and drawings.

About the author

Olivia 'Liv' Sisson is a forager and fungi enthusiast who has been enchanted by the minutiae of nature for as long as she can remember. Liv was raised in Virginia, USA, and first learned how to spot small wonders in the Blue Ridge Mountains. She studied geology and art before moving to Ōtautahi Christchurch, where she works for New Zealand Trade and Enterprise and leads foraging tours in addition to her work as a writer. Liv has spent countless hours exploring Aotearoa with a curious eye for the flora, fauna and fungi that define this special place. Her writing appears in *The Spinoff*, *Cuisine* and *Stuff*, she has featured on RNZ, and she loves how fungi-spotting micro-adventures introduce wonder and awe into her everyday.

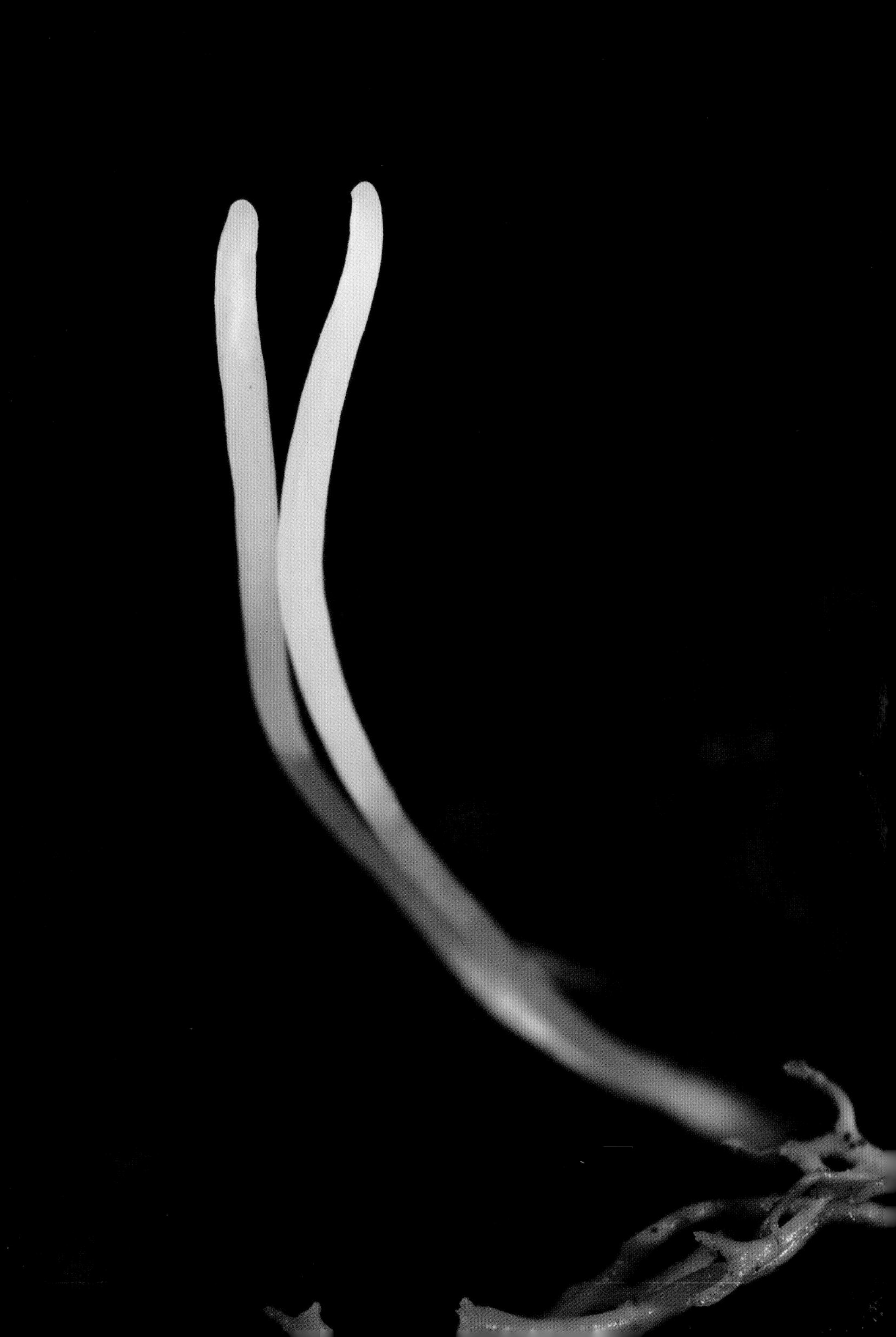